T0283594

Environmental Biochemistry

Environmental Biochemistry

Edited by
Erik Hamilton

Environmental Biochemistry
Edited by Erik Hamilton
ISBN: 978-1-63549-110-4 (Hardback)

Published by Larsen and Keller Education,
5 Penn Plaza,
19th Floor,
New York, NY 10001, USA

Cataloging-in-Publication Data

Environmental biochemistry / edited by Erik Hamilton.
p. cm.
Includes bibliographical references and index.
ISBN 978-1-63549-110-4
1. Biochemistry--Environmental aspects. 2. Environmental chemistry.
I. Hamilton, Erik.
TD193 .E58 2017
577.14--dc23

The publisher's policy is to use permanent paper from mills that operate a sustainable forestry policy. Furthermore, the publisher ensures that the text paper and cover boards used have met acceptable environmental accreditation standards.

Printed and bound in the United States of America.

For more information regarding Larsen and Keller Education and its products, please visit the publisher's website www.larsen-keller.com

Table of Contents

Preface

Environmental biochemistry is a part of environmental chemistry, which is the study of the various chemical and biochemical processes occurring in nature. It includes subfields like soil chemistry, atmospheric chemistry and also, aquatic chemistry. This book attempts to understand the multiple branches that fall under the discipline of environmental biochemistry and how such concepts have practical applications. It is compiled in such a manner, that it will provide in-depth knowledge about the theory and practice of the subject. For someone with an interest and eye for detail, this text covers the most significant topics in the field of environmental biochemistry. This textbook is meant for students who are looking for an elaborate reference text on this area.

A short introduction to every chapter is written below to provide an overview of the content of the book:

Chapter 1 - The scientific study of chemicals and biochemicals that occur naturally is known as environmental chemistry. It is an interdisciplinary science that includes aquatic and soil chemistry. This chapter will provide an integrated understanding of environmental chemistry; **Chapter 2 -** Biochemical cycle is a pathway that is used by chemical substances in order to move through the biotic and abiotic components of the Earth. The important cycles of biogeochemical cycles are oxygen cycle, sulfur cycle, carbon cycle, phosphorus cycle and water cycle. This section is an overview of the subject matter incorporating all the major aspects of biogeochemistry; **Chapter 3 -** Soil chemistry is the study of the features of the soil. It is mainly affected by mineral composition and environmental factors. The major environmental issues with soil are soil contamination, desertification, land surface effects on climate and soil erosion. Soil chemistry is best understood in confluence with the major topics listed in the following section; **Chapter 4 -** Environmental monitoring is the process that takes place in order to monitor the quality of the environment. The aspects elucidated in the section are air quality index, freshwater environmental quality parameters, water quality, oxygen saturation etc. The aspects elucidated in the section are of vital importance, and provide a better understanding of environmental monitoring; **Chapter 5 -** Bioindicators are biological species whose functions reveal the qualitative status of the environment. A biological monitor is an organism that provides quantitative information and this information provides qualitative information regarding the environment. This chapter strategically encompasses and incorporates the major components and the main concepts of bioindictors, providing a complete understanding; **Chapter 6 -** Green chemistry is an important area of chemistry that focuses on designing products and also focuses on minimizing the use of hazardous substances. Some of the aspects explained are natural-gas processing, supercritical hydrolysis and condensation reaction. This chapter on green chemistry offers an overview on the subject matter.

Finally, I would like to thank my fellow scholars who gave constructive feedback and my family members who supported me at every step.

Editor

1

Introduction to Environmental Chemistry

The scientific study of chemicals and biochemicals that occur naturally is known as environmental chemistry. It is an interdisciplinary science that includes aquatic and soil chemistry. This chapter will provide an integrated understanding of environmental chemistry.

White bags filled with contaminated stones line the shore near an industrial oil spill in Raahe, Finland

Environmental chemistry is the scientific study of the chemical and biochemical phenomena that occur in natural places. It should not be confused with green chemistry, which seeks to reduce potential pollution at its source. It can be defined as the study of the sources, reactions, transport, effects, and fates of chemical species in the air, soil, and water environments; and the effect of human activity and biological activity on these. Environmental chemistry is an interdisciplinary science that includes atmospheric, aquatic and soil chemistry, as well as heavily relying on analytical chemistry and being related to environmental and other areas of science.

Environmental chemistry is the study of chemical processes occurring in the environment which are impacted by humankind's activities. These impacts may be felt on a local scale, through the presence of urban air pollutants or toxic substances arising from a chemical waste site, or on a global scale, through depletion of stratospheric ozone or global warming. The focus in our courses and research activities is upon developing a fundamental understanding of the nature of these chemical processes, so that humankind's activities can be accurately evaluated.

Environmental chemistry involves first understanding how the uncontaminated environment works, which chemicals in what concentrations are present naturally, and with what effects. Without this it would be impossible to accurately study the effects humans have on the environment through the release of chemicals.

Environmental chemists draw on a range of concepts from chemistry and various environmental sciences to assist in their study of what is happening to a chemical species in the environment. Important general concepts from chemistry include understanding chemical reactions and equations, solutions, units, sampling, and analytical techniques.

Contamination

A contaminant is a substance present in nature at a level higher than fixed levels or that would not otherwise be there. This may be due to human activity and bioactivety. The term contaminant is often used interchangeably with *pollutant*, which is a substance that has a detrimental impact on the surrounding environment. Whilst a contaminant is sometimes defined as a substance present in the environment as a result of human activity, but without harmful effects, it is sometimes the case that toxic or harmful effects from contamination only become apparent at a later date.

The "medium" (e.g. soil) or organism (e.g. fish) affected by the pollutant or contaminant is called a *receptor*, whilst a *sink* is a chemical medium or species that retains and interacts with the pollutant e.g. as carbon sink and its effects by microbes.

Environmental Indicators

Chemical measures of water quality include dissolved oxygen (DO), chemical oxygen demand (COD), biochemical oxygen demand (BOD), total dissolved solids (TDS), pH, nutrients (nitrates and phosphorus), heavy metals (including copper, zinc, cadmium, lead and mercury), and pesticides.

Applications

Environmental chemistry is used by the Environment Agency (in England and Wales), the United States Environmental Protection Agency, the Association of Public Analysts, and other environmental agencies and research bodies around the world to detect and identify the nature and source of pollutants. These can include:

- Heavy metal contamination of land by industry. These can then be transported into water bodies and be taken up by living organisms.

- Nutrients leaching from agricultural land into water courses, which can lead to algal blooms and eutrophication.
- Urban runoff of pollutants washing off impervious surfaces (roads, parking lots, and rooftops) during rain storms. Typical pollutants include gasoline, motor oil and other hydrocarbon compounds, metals, nutrients and sediment (soil).
- Organometallic compounds.

Methods

Quantitative chemical analysis is a key part of environmental chemistry, since it provides the data that frame most environmental studies.

Common analytical techniques used for quantitative determinations in environmental chemistry include classical wet chemistry, such as gravimetric, titrimetric and electrochemical methods. More sophisticated approaches are used in the determination of trace metals and organic compounds. Metals are commonly measured by atomic spectroscopy and mass spectrometry: Atomic Absorption Spectrophotometry (AAS) and Inductively Coupled Plasma Atomic Emission (ICP-AES) or Inductively Coupled Plasma Mass Spectrometric (ICP-MS) techniques. Organic compounds are commonly measured also using mass spectrometric methods, such as Gas chromatography-mass spectrometry (GC/MS) and Liquid chromatography-mass spectrometry (LC/MS). Tandem Mass spectrometry MS/MS and High Resolution/Accurate Mass spectrometry HR/AM offer sub part per trillion detection. Non-MS methods using GCs and LCs having universal or specific detectors are still staples in the arsenal of available analytical tools.

Other parameters often measured in environmental chemistry are radiochemicals. These are pollutants which emit radioactive materials, such as alpha and beta particles, posing danger to human health and the environment. Particle counters and Scintillation counters are most commonly used for these measurements. Bioassays and immunoassays are utilized for toxicity evaluations of chemical effects on various organisms. Polymerase Chain Reaction PCR is able to identify species of bacteria and other organisms through specific DNA and RNA gene isolation and amplification and is showing promise as a valuable technique for identifying environmental microbial contamination.

Published Analytical Methods

Peer-reviewed test methods have been published by government agencies and private research organizations. Approved published methods must be used when testing to demonstrate compliance with regulatory requirements.

References

- Harrison, R.M (edited by). Understanding Our Environment, An Introduction to Environmental Chemistry and Pollution, Third Edition. Royal Society of Chemistry. 1999. ISBN 0-85404-584-8
- Sigel, A. (2010). Sigel, H.; Sigel, R.K.O., eds. Organometallics in Environment and Toxicology. Metal Ions in Life Sciences. 7. Cambridge: RSC publishing. ISBN 978-1-84755-177-1.
- vanLoon, Gary W.; Duffy, Stephen J. (2000). Environmental Chemistry. Oxford: Oxford. p. 7. ISBN 0-19-856440-6.
- EPA methods under the Resource Conservation and Recovery Act (RCRA): "Test Methods for Evaluating Solid Waste, Physical/Chemical Methods." Document No. SW-846. February 2007.

2

Understanding Biogeochemistry

Biochemical cycle is a pathway that is used by chemical substances in order to move through the biotic and abiotic components of the Earth. The important cycles of biogeochemical cycles are oxygen cycle, sulfur cycle, carbon cycle, phosphorus cycle and water cycle. This section is an overview of the subject matter incorporating all the major aspects of biogeochemistry.

Biogeochemistry

Biogeochemistry is the scientific discipline that involves the study of the chemical, physical, geological, and biological processes and reactions that govern the composition of the natural environment (including the biosphere, the cryosphere, the hydrosphere, the pedosphere, the atmosphere, and the lithosphere). In particular, biogeochemistry is the study of the cycles of chemical elements, such as carbon and nitrogen, and their interactions with and incorporation into living things transported through earth scale biological systems in space through time. The field focuses on chemical cycles which are either driven by or influence biological activity. Particular emphasis is placed on the study of carbon, nitrogen, sulfur, and phosphorus cycles. Biogeochemistry is a systems science closely related to systems ecology.

History

Vladimir Vernadsky - Russian geochemist.

The founder of biogeochemistry was Ukrainian scientist Vladimir Vernadsky whose 1926 book *The Biosphere*, in the tradition of Mendeleev, formulated a physics of the earth as a living whole. Vernadsky distinguished three spheres, where a sphere was a concept similar to the concept of a phase-space. He observed that each sphere had its own laws of evolution, and that the higher spheres modified and dominated the lower:

1. Abiotic sphere - all the non-living energy and material processes
2. Biosphere - the life processes that live within the abiotic sphere
3. Nöesis or Nösphere - the sphere of the cognitive process of man

Human activities (e.g., agriculture and industry) modify the Biosphere and Abiotic sphere. In the contemporary environment, the amount of influence humans have on the other two spheres is comparable to a geological force.

Early Development

The American limnologist and geochemist G. Evelyn Hutchinson is credited with outlining the broad scope and principles of this new field. More recently, the basic elements of the discipline of biogeochemistry were restated and popularized by the British scientist and writer, James Lovelock, under the label of the *Gaia Hypothesis*. Lovelock emphasizes a concept that life processes regulate the Earth through feedback mechanisms to keep it habitable.

Research

There are biogeochemistry research groups in many universities around the world. Since this is a highly inter-disciplinary field, these are situated within a wide range of host disciplines including: atmospheric sciences, biology, ecology, geomicrobiology, environmental chemistry, geology, oceanography and soil science. These are often bracketed into larger disciplines such as earth science and environmental science.

Many researchers investigate the biogeochemical cycles of chemical elements such as carbon, oxygen, nitrogen, phosphorus and sulfur, as well as their stable isotopes. The cycles of trace elements such as the trace metals and the radionuclides are also studied. This research has obvious applications in the exploration for ore deposits and oil, and in remediation of environmental pollution.

Some important research fields for biogeochemistry include:

- modelling of natural systems
- soil and water acidification recovery processes
- eutrophication of surface waters
- carbon sequestration

- soil remediation
- global change
- climate change
- biogeochemical prospecting for ore deposits

Biogeochemical Cycle

In Earth science, a biogeochemical cycle or substance turnover or cycling of substances is a pathway by which a chemical substance moves through both the biotic (biosphere) and abiotic (lithosphere, atmosphere, and hydrosphere) components of Earth. A cycle is a series of change which comes back to the starting point and which can be repeated. Water, for example, is always recycled through the water cycle. The water undergoes evaporation, condensation, and precipitation, falling back to Earth. Elements, chemical compounds, and other forms of matter are passed from one organism to another and from one part of the biosphere to another through biogeochemical cycles.'

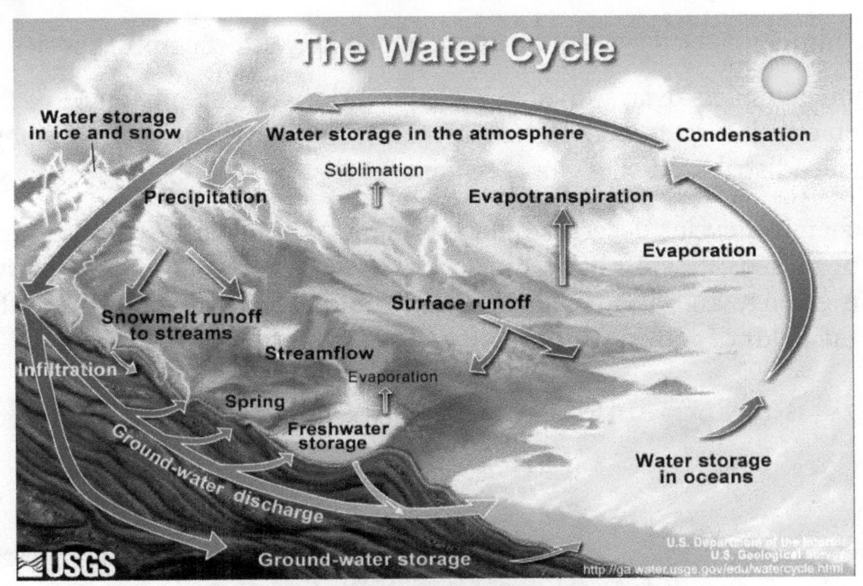

A commonly cited example is the water cycle.

The term "biogeochemical" tells us that biological, geological and chemical factors are all involved. The circulation of chemical nutrients like carbon, oxygen, nitrogen, phosphorus, calcium, and water etc. through the biological and physical world are known as biogeochemical cycles. In effect, the element is recycled, although in some cycles there may be places (called *reservoirs*) where the element is accumulated or held for a long period of time (such as an ocean or lake for water).

Important Cycles

The most well-known and important biogeochemical cycles, for example, include

- the carbon cycle,
- the nitrogen cycle,
- the oxygen cycle,
- the phosphorus cycle,
- the sulfur cycle,
- the water cycle,
- and the rock cycle.

There are many biogeochemical cycles that are currently being studied for the first time as climate change and human impacts are drastically changing the speed, intensity, and balance of these relatively unknown cycles. These newly studied biogeochemical cycles include

- the mercury cycle, and
- the human-caused cycle of atrazine, which may affect certain species.

Biogeochemical cycles always involve hot equilibrium states: a balance in the cycling of the element between compartments. However, overall balance may involve compartments distributed on a global scale.

As biogeochemical cycles describe the movements of substances on the entire globe, the study of these is inherently multidisciplinary. The carbon cycle may be related to research in ecology and atmospheric sciences. Biochemical dynamics would also be related to the fields of geology and pedology (soil study).

Important Biogeochemical Cycles

Oxygen Cycle

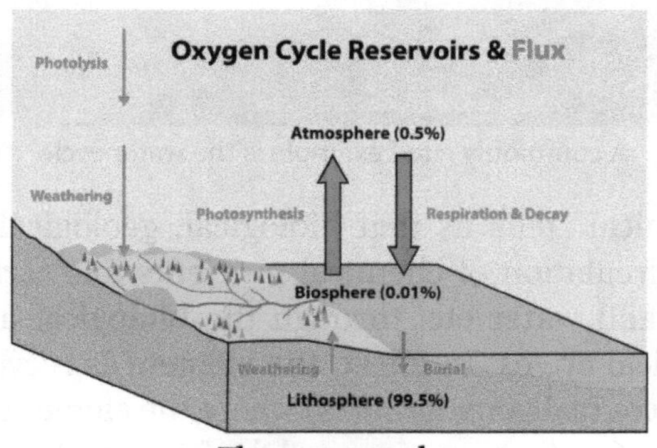

The oxygen cycle

The oxygen cycle is the biogeochemical cycle of oxygen within its three main reservoirs: the atmosphere (air), the total content of biological matter within the biosphere (the global sum of all ecosystems), and the Earth's crust. Failures in the oxygen cycle within the hydrosphere (the combined mass of water found on, under, and over the surface of planet Earth) can result in the development of hypoxic zones. The main driving factor of the oxygen cycle is photosynthesis, which is responsible for the modern Earth's atmosphere and life on earth.

Reservoirs

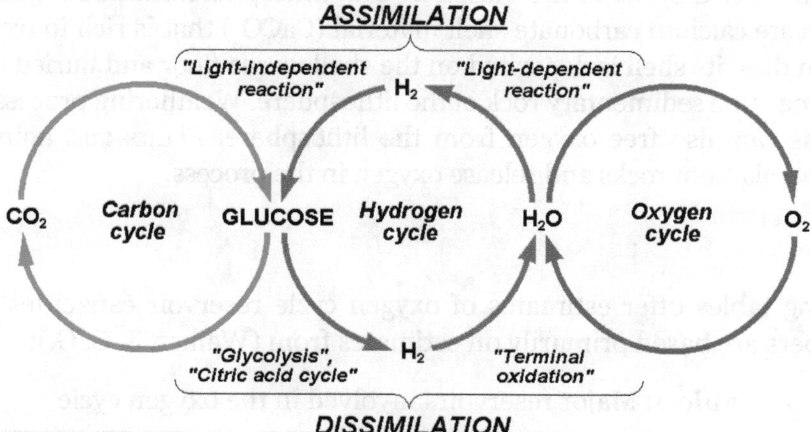

Interconnection between carbon, hydrogen and oxygen cycle in metabolism of photosynthesizing plants

By far the largest reservoir of Earth's oxygen is within the silicate and oxide minerals of the crust and mantle (99.5%). Only a small portion has been released as free oxygen to the biosphere (0.01%) and atmosphere (0.36%). The main source of atmospheric free oxygen is photosynthesis, which produces sugars and free oxygen from carbon dioxide and water:

$$6\ CO_2 + 6H_2O + energy \rightarrow C_6H_{12}O_6 + 6\ O_2$$

Photosynthesizing organisms include the plant life of the land areas as well as the phytoplankton of the oceans. The tiny marine cyanobacterium Prochlorococcus was discovered in 1986 and accounts for more than half of the photosynthesis of the open ocean.

An additional source of atmospheric free oxygen comes from photolysis, whereby high-energy ultraviolet radiation breaks down atmospheric water and nitrous oxide into component atoms. The free H and N atoms escape into space, leaving O_2 in the atmosphere:

$$2\ H_2O + energy \rightarrow 4\ H + O_2$$

$$2\ N_2O + energy \rightarrow 4\ N + O_2$$

The main way free oxygen is lost from the atmosphere is via respiration and decay, mechanisms in which animal life and bacteria consume oxygen and release carbon dioxide.

The lithosphere also consumes free oxygen by chemical weathering and surface reactions. An example of surface weathering chemistry is formation of iron oxides (rust):

$$4\,FeO + O_2 \rightarrow 2\,Fe_2O_3$$

Oxygen is also cycled between the biosphere and lithosphere. Marine organisms in the biosphere create calcium carbonate shell material ($CaCO_3$) that is rich in oxygen. When the organism dies, its shell is deposited on the shallow sea floor and buried over time to create the limestone sedimentary rock of the lithosphere. Weathering processes initiated by organisms can also free oxygen from the lithosphere. Plants and animals extract nutrient minerals from rocks and release oxygen in the process.

Capacities and Fluxes

The following tables offer estimates of oxygen cycle reservoir capacities and fluxes. These numbers are based primarily on estimates from (Walker, J. C. G.):

Table 1: Major reservoirs involved in the oxygen cycle

Reservoir	Capacity (kg O_2)	Flux in/out (kg O_2 per year)	Residence time (years)
Atmosphere	1.4×10^{18}	3×10^{14}	4500
Biosphere	1.6×10^{16}	3×10^{14}	50
Lithosphere	2.9×10^{20}	6×10^{11}	500000000

Table 2: Annual gain and loss of atmospheric oxygen (Units of 10^{10} kg O_2 per year)

Photosynthesis (land) Photosynthesis (ocean) Photolysis of N_2O Photolysis of H_2O	16,500 13,500 1.3 0.03
Total gains	~ 30,000
Losses - respiration and decay	

Aerobic respiration	23,000
Microbial oxidation	5,100
Combustion of fossil fuel (anthropogenic)	1,200
Photochemical oxidation	600
Fixation of N_2 by lightning	12
Fixation of N_2 by industry (anthropogenic)	10
Oxidation of volcanic gases	5
Losses - weathering	
Chemical weathering	50
Surface reaction of O_3	12
Total losses	~ 30,000

Ozone

The presence of atmospheric oxygen has led to the formation of ozone (O_3) and the ozone layer within the stratosphere:

$$O_2 + \text{uv light} \rightarrow 2O \qquad (\lambda \lesssim 200\,\text{nm})$$

$$O + O_2 \rightarrow O_3$$

The ozone layer is extremely important to modern life as it absorbs harmful ultraviolet radiation:

$$O_3 + \text{uv light} \rightarrow O_2 + O \qquad (\lambda \lesssim 300\,\text{nm})$$

Sulfur Cycle

The sulfur cycle is the collection of processes by which sulfur moves to and from minerals (including the waterways) and living systems. Such biogeochemical cycles are important in geology because they affect many minerals. Biochemical cycles are also important for life because sulfur is an essential element, being a constituent of many proteins and cofactors.

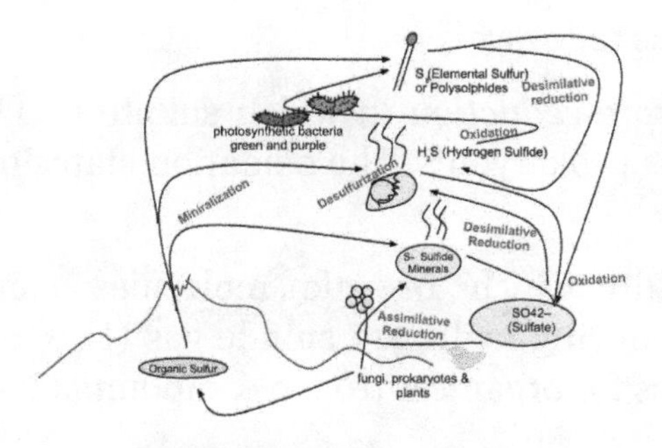

The Sulfur cycle (in general)

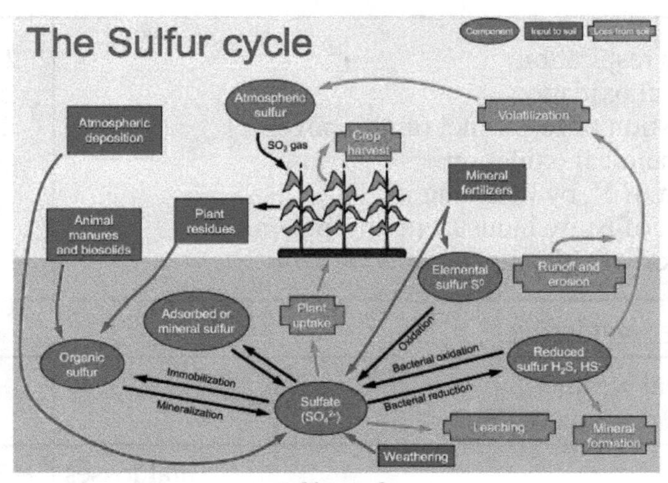

Sulfur cycle

Steps of the sulfur cycle are:

- Mineralization of organic sulfur into inorganic forms, such as hydrogen sulfide (H_2S), elemental sulfur, as well as sulfide minerals.
- Oxidation of hydrogen sulfide, sulfide, and elemental sulfur (S) to sulfate (SO_4^{2-}).
- Reduction of sulfate to sulfide.
- Incorporation of sulfide into organic compounds (including metal-containing derivatives).

Structure of 3'-phosphoadenosine-5'-phosphosulfate, a key intermediate in the sulfur cycle.

These are often termed as follows:

Assimilative sulfate reduction in which sulfate (SO_4^{2-}) is reduced by plants, fungi and various prokaryotes. The oxidation states of sulfur are +6 in sulfate and −2 in R–SH.

Desulfurization in which organic molecules containing sulfur can be desulfurized, producing hydrogen sulfide gas (H_2S, oxidation state = −2). An analogous process for organic nitrogen compounds is deamination.

Oxidation of hydrogen sulfide produces elemental sulfur (S_8), oxidation state = 0. This reaction occurs in the photosynthetic green and purple sulfur bacteria and some chemolithotrophs. Often the elemental sulfur is stored as polysulfides.

Oxidation of elemental sulfur by sulfur oxidizers produces sulfate.

Dissimilative sulfur reduction in which elemental sulfur can be reduced to hydrogen sulfide.

Dissimilative sulfate reduction in which sulfate reducers generate hydrogen sulfide from sulfate.

Sulfur Sources and Sinks

Sulfur is found in oxidation states ranging from +6 in SO_4^{2-} to -2 in sulfides. Thus elemental sulfur can either give or receive electrons depending on its environment. Minerals such as pyrite (FeS_2) comprise the original pool of sulfur on earth. Owing to the sulfur cycle, the amount of mobile sulfur has been continuously increasing through volcanic activity as well as weathering of the crust in an oxygenated atmosphere. Earth's main sulfur sink is the oceans SO_4^{2-}, where it is the major oxidizing agent.

When SO_4^{2-} is assimilated by organisms, it is reduced and converted to organic sulfur, which is an essential component of proteins. However, the biosphere does not act as a major sink for sulfur, instead the majority of sulfur is found in seawater or sedimentary rocks especially pyrite rich shales and evaporite rocks (anhydrite and baryte). The amount of sulfate in the oceans is controlled by three major processes:

1. input from rivers
2. sulfate reduction and sulfide re-oxidation on continental shelves and slopes
3. burial of anhydrite and pyrite in the oceanic crust.

There is no significant amount of sulfur held in the atmosphere with all of it coming from either sea spray or windblown sulfur rich dust, neither of which is long lived in the atmosphere. In recent times the large annual input of sulfur from the burning of coal and other fossil fuels adds a substantial amount SO_2 which acts as an air pollutant. In the geologic past, igneous intrusions into coal measures have caused large scale burning of these measures, and consequential release of sulfur to the atmosphere. This has led to substantial disruption to the climate system, and is one of the proposed causes of the great dying.

Dimethylsulfide [$(CH_3)_2S$ or DMS] is produced by the decomposition of dimethylsulfoniopropionate (DMSP) from dying phytoplankton cells in the shallow levels of the ocean, and is the major biogenic gas emitted from the sea, where it is responsible for the distinctive "smell of the sea" along coastlines. DMS is the largest natural source of sulfur gas, but still only has a residence time of about one day in the atmosphere and a majority of it is redeposited in the oceans rather than making it to land. However, it is a significant factor in the climate system, as it is involved in the formation of clouds.

Biologically and Thermochemically Driven Sulfate Reduction

Sulfur can be reduced both biologically and thermochemically. Dissimilarity sulfate reduction has two different definitions:

1. the microbial process that converts sulfate to sulfide for energy gain, and
2. a set of forward and reverse pathways that progress from the uptake and release of sulfate by the cell to its conversion to various sulfur intermediates, and ultimately to sulfide which is released from the cell.

Sulfide and thiosulfate are the most abundant reduced inorganic sulfur species in the environments and are converted to sulfate, primarily by bacterial action, in the oxidative half of the sulfur cycle. Bacterial sulfate reduction (BSR) can only occur at temperature from 0 up to 60–80 °C because above that temperature almost all sulfate-reducing microbes can no longer metabolize. Few microbes can form H_2S at higher temperatures but appear to be very rare and do not metabolize in settings where normal bacterial sulfate reduction is occurring. BSR is geologically instantaneous happening on the order of hundreds to thousands of years. Thermochemical sulfate reduction (TSR) occurs at much higher temperatures (160–180 °C) and over longer time intervals, several tens of thousands to a few million years.

The main difference between these two reactions is obvious, one is organically driven and the other is chemically driven. Therefore, the temperature for thermochemical sulfate reduction is much higher due to the activation energy required to reduce sulfate. Bacterial sulfate reductions requires lower temperatures because the sulfur reducing bacteria can only live at relatively low temperature (below 60 °C). BSR also requires a relatively open system; otherwise the bacteria will poison themselves when the sulfate levels rise above 5–10%.

The organic reactants involved in BSR are organic acids which are distinctive from the organic reactants needed for TSR. In both cases sulfate is usually derived from the dissolution of gypsum or taken directly out of the seawater. The factors that control whether BSR or TSR will occur are temperature, which is generally a product of depth, with BSR occurring in shallower levels than TSR. Both can occur within the oil window. Their solid products are similar but can be distinguished from one another petrographically, due to their differing crystal sizes, shapes and reflectivity.

$\delta^{34}S$

Although 25 isotopes are known for sulfur, only four are stable and of geochemical importance. Of those four, two (^{32}S, light and ^{34}S, heavy) comprise (99.22%) of S on

Earth. The vast majority (95.02%) of S occurs as ^{32}S with only 4.21% in ^{34}S. The ratio of these two isotopes is fixed in our solar system and has been since its formation. The bulk Earth sulfur isotopic ratio is thought to be the same as the ratio of 22.22 measured from the Canyon Diablo troilite (CDT), a meteorite. That ratio is accepted as the international standard and is therefore set at δ0.00. Deviation from 0.00 is expressed as the $\delta^{34}S$ which is a ratio in per mill (‰). Positive values correlate to increased levels of ^{34}S, whereas negative values correlate with greater ^{32}S in a sample.

Formation of sulfur minerals through non-biogenic processes does not substantially differentiate between the light and heavy isotopes, therefore sulfur values in gypsum or baryte should be the same as the overall ratio in the water column at their time of precipitation. Sulfate reduction through biologic activity strongly differentiates between the two isotopes because of the more rapid enzymic reaction with ^{32}S. Sulfate metabolism results in an isotopic depletion of -18‰, and repeated cycles of oxidation and reduction can result in values up to -50 ‰. Average present day seawater values of $\delta^{34}S$ are on the order of +21‰.

Throughout geologic history the sulfur cycle and the isotopic ratios have coevolved with the biosphere becoming overall more negative with the increases in biologically driven sulfate reduction, but also show substantial positive excursion. In general positive excursions in the sulfur isotopes mean that there is an excess of pyrite deposition rather than oxidation of sulfide minerals exposed on land.

Evolution of the Sulfur Cycle

The isotopic composition of sedimentary sulfides provides primary information on the evolution of the sulfur cycle.

The total inventory of sulfur compounds on the surface of the Earth (nearly 10^{22} g S) represents the total outgassing of sulfur through geologic time. Rocks analyzed for sulfur content are generally organic-rich shales meaning they are likely controlled by biogenic sulfur reduction. Average seawater curves are generated from evaporites deposited throughout geologic time because again, since they do not discriminate between the heavy and light sulfur isotopes, they should mimic the ocean composition at the time of deposition.

4.6 billion years ago (Ga) the Earth formed and had a theoretical $\delta^{34}S$ value of 0. Since there was no biologic activity on early Earth there would be no isotopic fractionation. All sulfur in the atmosphere would be released during volcanic eruptions. When the oceans condensed on Earth, the atmosphere was essentially swept clean of sulfur gases, owing to their high solubility in water. Throughout the majority of the Archean (4.6–2.5 Ga) most systems appeared to be sulfate-limited. Some small Archean evaporite deposits require that at least locally elevated concentrations (possibly due to local volcanic activity) of sulfate existed in order for them to be supersaturated and precipitate out of solution.

3.8–3.6 Ga marks the beginning of the exposed geologic record because this is the age of the oldest rocks on Earth. Metasedimentary rocks from this time still have an isotopic value of 0 because the biosphere was not developed enough (possibly at all) to fractionate sulfur.

3.5 Ga anoxyogenic photosynthesis is established and provides a weak source of sulfate to the global ocean with sulfate concentrations incredibly low the $\delta^{34}S$ is still basically 0. Shortly after, at 3.4 Ga the first evidence for minimal fractionation in evaporitic sulfate in association with magmatically derived sulfides can be seen in the rock record. This fractionation shows possible evidence for anoxygenic phototrophic bacteria.

2.8 Ga marks the first evidence for oxygen production through photosynthesis. This is important because there cannot be sulfur oxidation without oxygen in the atmosphere. This exemplifies the coevolution of the oxygen and sulfur cycles as well as the biosphere.

2.7–2.5 Ga is the age of the oldest sedimentary rocks to have a depleted $\delta\ ^{34}S$ which provide the first compelling evidence for sulfate reduction.

2.3 Ga sulfate increases to more than 1 mM; this increase in sulfate is coincident with the "Great Oxygenation Event", when redox conditions on Earth's surface are thought by most workers to have shifted fundamentally from reducing to oxidizing. This shift would have led to an incredible increase in sulfate weathering which would have led to an increase in sulfate in the oceans. The large isotopic fractionations that would likely be associated with bacteria reduction are produced for the first time. Although there was a distinct rise in seawater sulfate at this time it was likely still only less than 5–15% of present-day levels.

At 1.8 Ga, Banded iron formations (BIF) are common sedimentary rocks throughout the Archean and Paleoproterozoic; their disappearance marks a distinct shift in the chemistry of ocean water. BIFs have alternating layers of iron oxides and chert. BIFs only form if the water is be allowed to supersaturate in dissolved iron (Fe^{2+}) meaning there cannot be free oxygen or sulfur in the water column because it would form Fe^{3+} (rust) or pyrite and precipitate out of solution. Following this supersaturation, the water must become oxygenated in order for the ferric rich bands to precipitate it must still be sulfur poor otherwise pyrite would form instead of Fe^{3+}. It has been hypothesized that BIFs formed during to the initial evolution of photosynthetic organisms that had phases of population growth, causing over production of oxygen. Due to this over production they would poison themselves causing a mass die off, which would cut off the source of oxygen and produce a large amount of CO_2 through the decomposition of their bodies, allowing for another bacterial bloom. After 1.8 Ga sulfate concentrations were sufficient to increase rates of sulfate reduction to greater than the delivery flux of iron to the oceans.

Along with the disappearance of BIF, the end of the Paleoproterozoic also marks the first large scale sedimentary exhalative deposits showing a link between mineralization and a likely increase in the amount of sulfate in sea water. In the Paleoproterozoic the sulfate in seawater had increased to an amount greater than in the Archean, but was still lower than present day values. The sulfate levels in the Proterozoic also act as proxies for atmospheric oxygen because sulfate is produced mostly through weathering of the continents in the presence of oxygen. The low levels in the Proterozoic simply imply that levels of atmospheric oxygen fell between the abundances of the Phanerozoic and the deficiencies of the Archean.

750 million years ago (Ma) there is a renewed deposition of BIF which marks a significant change in ocean chemistry. This was likely due to snowball earth episodes where the entire globe including the oceans was covered in a layer of ice cutting off oxygenation. In the late Neoproterozoic high carbon burial rates increased the atmospheric oxygen level to >10% of its present-day value. In the Latest Neoproterozoic another major oxidizing event occurred on Earth's surface that resulted in an oxic deep ocean and possibly allowed for the appearance of multicellular life.

During the last 600 million years, seawater SO_4 has varied between +10 and +30‰ in $\delta^{34}S$, with an average value close to that of today. This coincides with atmospheric O levels reaching something close to modern values around the Precambrian–Cambrian boundary.

Over a shorter time scale (ten million years) changes in the sulfur cycle are easier to observe and can be even better constrained with oxygen isotopes. Oxygen is continually incorporated into the sulfur cycle through sulfate oxidation and then released when that sulfate is reduced once again. Since different sulfate sources within the ocean have distinct oxygen isotopic values it may be possible to use oxygen to trace the sulfur cycle. Biological sulfate reduction preferentially selects lighter oxygen isotopes for the same reason that lighter sulfur isotopes are preferred. By studying oxygen isotopes in ocean sediments over the last 10 million years were able to better constrain the sulfur concentrations in sea water through that same time. They found that the sea level changes due to Pliocene and Pleistocene glacial cycles changed the area of continental shelves which then disrupted the sulfur processing, lowering the concentration of sulfate in the sea water. This was a drastic change as compared to preglacial times before 2 million years ago.

Economic Importance

Sulfur is intimately involved in production of fossil fuels and a majority of metal deposits because of its ability to act as an oxidizing or reducing agent. The vast majority of the major mineral deposits on Earth contain a substantial amount of sulfur including, but not limited to: sedimentary exhalative deposits(SEDEX), Mississippi Valley-Type (MVT) and copper porphyry deposits. Iron sulfides, galena and sphalerite will form as by-products of hydrogen sulfide generation, as long as the respective transition or base metals are present or transported to a sulfate reduction site. If the system runs out of reactive hydrocarbons economically viable elemental sulfur deposits may form. Sulfur also acts as a reducing agent in many natural gas reservoirs and generally ore forming fluids have a close relationship with ancient hydrocarbon seeps or vents.

Important sources of sulfur in ore deposits are generally deep-seated, but they can also come from local country rocks, sea water, or marine evaporites. The presence or absence of sulfur is one of the limiting factors on both the concentration of precious metals and its precipitation from solution. pH, temperature and especially redox states determine whether sulfides will precipitate. Most sulfide brines will remain in concentration until they reach reducing conditions, a higher pH or lower temperatures.

Ore fluids are generally linked to metal rich waters that have been heated within a sedimentary basin under the elevated thermal conditions typically in extensional tectonic settings. The redox conditions of the basin lithologies exert an important control on the redox state of the metal-transporting fluids and deposits can form from both oxidizing and reducing fluids. Metal-rich ore fluids tend to be by necessity comparatively sulfide deficient, so a substantial portion of the sulfide must be supplied from another source at the site of mineralization. Bacterial reduction of seawater sulfate or a euxinic (anoxic and H_2S-containing) water column is a necessary source of that sulfide. When present, the $\delta^{34}S$ values of barite are generally consistent with a seawater sulfate source, suggesting barite formation by reaction between hydrothermal barium and sulfate in ambient seawater.

Once fossil fuels or precious metals are discovered and either burned or milled, the sulfur become a waste product which must be dealt with properly or it can become a pollutant. There has been a great increase in the amount of sulfur in our present day atmosphere because of the burning of fossil fuels. Sulfur acts as a pollutant and an economic resource at the same time.

Human Impact

Human activities have a major effect on the global sulfur cycle. The burning of coal, natural gas, and other fossil fuels has greatly increased the amount of S in the atmosphere and ocean and depleted the sedimentary rock sink. Without human impact sulfur would stay tied up in rocks for millions of years until it was uplifted through tectonic events and then released through erosion and weathering processes. Instead it is being drilled, pumped and burned at a steadily increasing rate. Over the most polluted areas there has been a 30-fold increase in sulfate deposition.

Although the sulfur curve shows shifts between net sulfur oxidation and net sulfur reduction in the geologic past, the magnitude of the current human impact is probably unprecedented in the geologic record. Human activities greatly increase the flux of sulfur to the atmosphere, some of which is transported globally. Humans are mining coal and extracting petroleum from the Earth's crust at a rate that mobilizes 150×10^{12} gS/yr, which is more than double the rate of 100 years ago. The result of human impact on these processes is to increase the pool of oxidized sulfur (SO_4) in the global cycle, at the expense of the storage of reduced sulfur in the Earth's crust. Therefore, human activities do not cause a major change in the global pools of S, but they do produce massive changes in the annual flux of S through the atmosphere.

When SO_2 is emitted as an air pollutant, it forms sulfuric acid through reactions with water in the atmosphere. Once the acid is completely dissociated in water the pH can drop to 4.3 or lower causing damage to both man-made and natural systems. According to the EPA, acid rain is a broad term referring to a mixture of wet and dry deposition (deposited material) from the atmosphere containing higher than normal amounts of nitric and sulfuric acids. Distilled water (water without any dissolved constituents), which contains no carbon dioxide, has a neutral pH of 7. Rain naturally has a slightly acidic pH of 5.6, because carbon dioxide and water in the air react together to form carbonic acid, a very weak acid. Around Washington, D.C., however, the average rain pH is between 4.2 and 4.4. Since pH is on a log scale dropping by 1 (the difference between normal rain water and acid rain) has a dramatic effect on the strength of the acid. In the United States, roughly 2/3 of all SO_2 and 1/4 of all NO_3 come from electric power generation that relies on burning fossil fuels, like coal.

Carbon Cycle

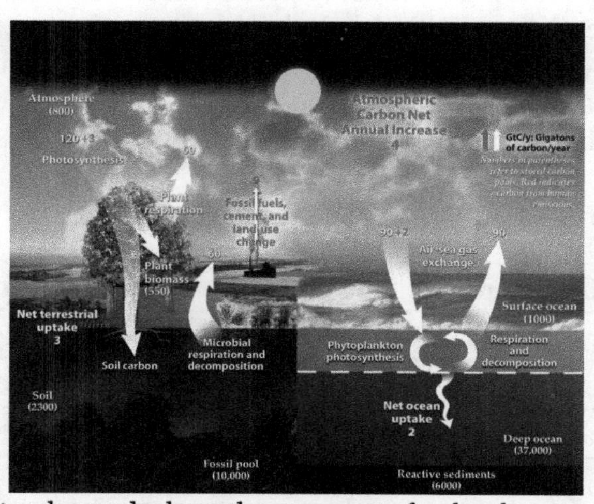

This diagram of the fast carbon cycle shows the movement of carbon between land, atmosphere, and oceans in billions of tons per year. Yellow numbers are natural fluxes, red are human contributions, white indicate stored carbon. Note this diagram does not account for volcanic and tectonic activity, which also sequesters and releases carbon.

The carbon cycle is the biogeochemical cycle by which carbon is exchanged among the biosphere, pedosphere, geosphere, hydrosphere, and atmosphere of the Earth. Along with the nitrogen cycle and the water cycle, the carbon cycle comprises a sequence of events that are key to making the Earth capable of sustaining life; it describes the movement of carbon as it is recycled and reused throughout the biosphere, including carbon sinks.

The *global carbon budget* is the balance of the exchanges (incomes and losses) of carbon between the carbon reservoirs or between one specific loop (e.g., atmosphere <-> biosphere) of the carbon cycle. An examination of the carbon budget of a pool or reservoir can provide information about whether the pool or reservoir is functioning as a source or sink for carbon dioxide. The carbon cycle was initially discovered by Joseph Priestley and Antoine Lavoisier, and popularized by Humphry Davy.

Global Climate

Carbon-based molecules are crucial for life on Earth, because it is the main component of biological compounds. Carbon is also a major component of many minerals. Carbon also exists in various forms in the atmosphere. Carbon dioxide (CO_2) is partly responsible for the greenhouse effect and is the most important human-contributed greenhouse gas.

In the past two centuries, human activities have seriously altered the global carbon cycle, most significantly in the atmosphere. Although carbon dioxide levels have changed naturally over the past several thousand years, human emissions of carbon dioxide into the atmosphere exceed natural fluctuations. Changes in the amount of atmospheric CO_2 are considerably altering weather patterns and indirectly influencing oceanic chemistry. Current carbon dioxide levels in the atmosphere exceed measurements from the last 420,000 years and levels are rising faster than ever recorded, making it of critical importance to better understand how the carbon cycle works and what its effects are on the global climate.

Main Components

Carbon pools in the major reservoirs on earth.

Pool	Quantity (gigatons)
Atmosphere	720
Oceans (total)	38,400
Total inorganic	*37,400*
Total organic	1,000
Surface layer	670
Deep layer	36,730
Lithosphere	
Sedimentary carbonates	> 60,000,000
Kerogens	15,000,000
Terrestrial biosphere (total)	2,000

Living biomass	600 - 1,000
Dead biomass	1,200
Aquatic biosphere	1 - 2
Fossil fuels (total)	4,130
Coal	3,510
Oil	230
Gas	140
Other (peat)	250

The global carbon cycle is now usually divided into the following major reservoirs of carbon interconnected by pathways of exchange:

- The atmosphere
- The terrestrial biosphere
- The oceans, including dissolved inorganic carbon and living and non-living marine biota
- The sediments, including fossil fuels, fresh water systems and non-living organic material.
- The Earth's interior, carbon from the Earth's mantle and crust. These carbon stores interact with the other components through geological processes

The carbon exchanges between reservoirs occur as the result of various chemical, physical, geological, and biological processes. The ocean contains the largest active pool of carbon near the surface of the Earth. The natural flows of carbon between the atmosphere, ocean, terrestrial ecosystems, and sediments is fairly balanced, so that carbon levels would be roughly stable without human influence.

Atmosphere

Carbon in the Earth's atmosphere exists in two main forms: carbon dioxide and methane. Both of these gases absorb and retain heat in the atmosphere and are partially responsible for the greenhouse effect. Methane produces a larger greenhouse effect per volume as compared to carbon dioxide, but it exists in much lower concentrations and is more short-lived than carbon dioxide, making carbon dioxide the more important greenhouse gas of the two.

Carbon dioxide leaves the atmosphere through photosynthesis, thus entering the terrestrial and oceanic biospheres. Carbon dioxide also dissolves directly from the atmosphere into bodies of water (oceans, lakes, etc.), as well as dissolving in precipitation as raindrops fall through the atmosphere. When dissolved in water, carbon dioxide reacts with water molecules and forms carbonic acid, which contributes to ocean acidity. It can then be absorbed by rocks through weathering. It also can acidify other surfaces it touches or be washed into the ocean.

Epiphytes on electric wires. This kind of plant takes both CO_2 and water from the atmosphere for living and growing.

Human activities over the past two centuries have significantly increased the amount of carbon in the atmosphere, mainly in the form of carbon dioxide, both by modifying ecosystems' ability to extract carbon dioxide from the atmosphere and by emitting it directly, e.g., by burning fossil fuels and manufacturing concrete.

Terrestrial Biosphere

A portable soil respiration system measuring soil CO_2 flux

The terrestrial biosphere includes the organic carbon in all land-living organisms, both alive and dead, as well as carbon stored in soils. About 500 gigatons of carbon are stored above ground in plants and other living organisms, while soil holds approximately 1,500 gigatons of carbon. Most carbon in the terrestrial biosphere is organic carbon, while about a third of soil carbon is stored in inorganic forms, such as calcium carbonate. Organic carbon is a major component of all organisms living on earth. Autotrophs extract it from the air in the form of carbon dioxide, converting it into organic carbon, while heterotrophs receive carbon by consuming other organisms.

Because carbon uptake in the terrestrial biosphere is dependent on biotic factors, it follows a diurnal and seasonal cycle. In CO_2 measurements, this feature is apparent in the Keeling curve. It is strongest in the northern hemisphere, because this hemisphere has more land mass than the southern hemisphere and thus more room for ecosystems to absorb and emit carbon.

Carbon leaves the terrestrial biosphere in several ways and on different time scales. The combustion or respiration of organic carbon releases it rapidly into the atmosphere. It can also be exported into the oceans through rivers or remain sequestered in soils in the form of inert carbon. Carbon stored in soil can remain there for up to thousands of years before being washed into rivers by erosion or released into the atmosphere through soil respiration. Between 1989 and 2008 soil respiration increased by about 0.1% per year. In 2008, the global total of CO_2 released from the soil reached roughly 98 billion tonnes, about 10 times more carbon than humans are now putting into the atmosphere each year by burning fossil fuel. There are a few plausible explanations for this trend, but the most likely explanation is that increasing temperatures have increased rates of decomposition of soil organic matter, which has increased the flow of CO_2. The length of carbon sequestering in soil is dependent on local climatic conditions and thus changes in the course of climate change. From pre-industrial era to 2010, the terrestrial biosphere represented a net source of atmospheric CO_2 prior to 1940, switching subsequently to a net sink.

Oceans

Oceans contain the greatest quantity of actively cycled carbon in this world and are second only to the lithosphere in the amount of carbon they store. The oceans' surface layer holds large amounts of dissolved inorganic carbon that is exchanged rapidly with the atmosphere. The deep layer's concentration of dissolved inorganic carbon (DIC) is about 15% higher than that of the surface layer. DIC is stored in the deep layer for much longer periods of time. Thermohaline circulation exchanges carbon between these two layers.

Carbon enters the ocean mainly through the dissolution of atmospheric carbon dioxide, which is converted into carbonate. It can also enter the oceans through rivers as dissolved organic carbon. It is converted by organisms into organic carbon through photosynthesis and can either be exchanged throughout the food chain or precipitated into the ocean's deeper, more carbon rich layers as dead soft tissue or in shells as calcium carbonate. It circulates in this layer for long periods of time before either being deposited as sediment or, eventually, returned to the surface waters through thermohaline circulation.

Oceanic absorption of CO_2 is one of the most important forms of carbon sequestering limiting the human-caused rise of carbon dioxide in the atmosphere. However, this process is limited by a number of factors. Because the rate of CO_2 dissolution in the ocean is dependent on the weathering of rocks and this process takes place slower than current rates of human greenhouse gas emissions, ocean CO_2 uptake will decrease in the future. CO_2 absorption also makes water more acidic, which affects ocean biosystems. The projected rate of increasing oceanic acidity could slow the biological precipitation of calcium carbonates, thus decreasing the ocean's capacity to absorb carbon dioxide.

Geological Carbon Cycle

The geologic component of the carbon cycle operates slowly in comparison to the other parts of the global carbon cycle. It is one of the most important determinants of the amount of carbon in the atmosphere, and thus of global temperatures.

Most of the earth's carbon is stored inertly in the earth's lithosphere. Much of the carbon stored in the earth's mantle was stored there when the earth formed. Some of it was deposited in the form of organic carbon from the biosphere. Of the carbon stored in the geosphere, about 80% is limestone and its derivatives, which form from the sedimentation of calcium carbonate stored in the shells of marine organisms. The remaining 20% is stored as kerogens formed through the sedimentation and burial of terrestrial organisms under high heat and pressure. Organic carbon stored in the geosphere can remain there for millions of years.

Carbon can leave the geosphere in several ways. Carbon dioxide is released during the metamorphosis of carbonate rocks when they are subducted into the earth's mantle. This carbon dioxide can be released into the atmosphere and ocean through volcanoes and hotspots. It can also be removed by humans through the direct extraction of kerogens in the form of fossil fuels. After extraction, fossil fuels are burned to release energy, thus emitting the carbon they store into the atmosphere.

Human Influence

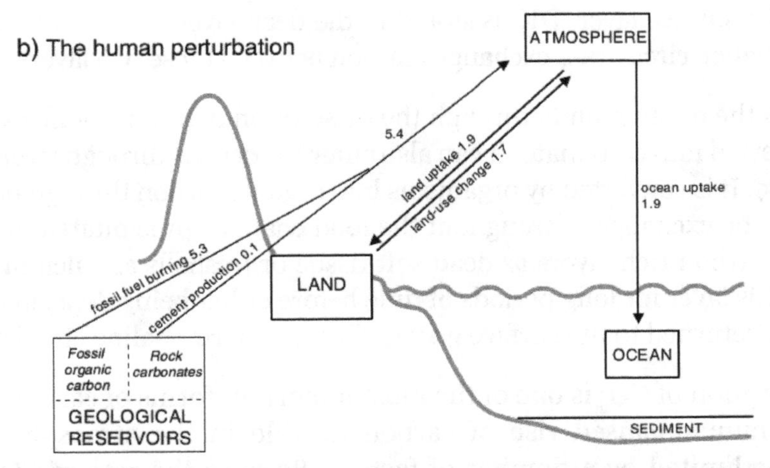

Human activity since the industrial era has changed the balance in the natural carbon cycle. Units are in gigatons.

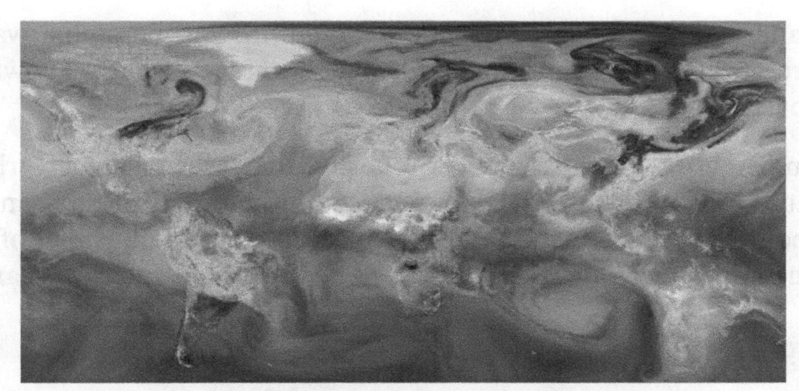

CO_2 in Earth's atmosphere if *half* of global-warming emissions are *not* absorbed. (NASA computer simulation).

Since the industrial revolution, human activity has modified the carbon cycle by changing its components' functions and directly adding carbon to the atmosphere.

The largest human impact on the carbon cycle is through direct emissions from burning fossil fuels, which transfers carbon from the geosphere into the atmosphere. The rest of this increase is caused mostly by changes in land-use, particularly deforestation.

Another direct human impact on the carbon cycle is the chemical process of calcination of limestone for clinker production, which releases CO_2. Clinker is an industrial precursor of cement.

Humans also influence the carbon cycle indirectly by changing the terrestrial and oceanic biosphere. Over the past several centuries, direct and indirect human-caused land use and land cover change (LUCC) has led to the loss of biodiversity, which lowers ecosystems' resilience to environmental stresses and decreases their ability to remove carbon from the atmosphere. More directly, it often leads to the release of carbon from terrestrial ecosystems into the atmosphere. Deforestation for agricultural purposes removes forests, which hold large amounts of carbon, and replaces them, generally with agricultural or urban areas. Both of these replacement land cover types store comparatively small amounts of carbon, so that the net product of the process is that more carbon stays in the atmosphere.

Other human-caused changes to the environment change ecosystems' productivity and their ability to remove carbon from the atmosphere. Air pollution, for example, damages plants and soils, while many agricultural and land use practices lead to higher erosion rates, washing carbon out of soils and decreasing plant productivity.

Humans also affect the oceanic carbon cycle. Current trends in climate change lead to higher ocean temperatures, thus modifying ecosystems. Also, acid rain and polluted runoff from agriculture and industry change the ocean's chemical composition. Such changes can have dramatic effects on highly sensitive ecosystems such as coral reefs, thus limiting the ocean's ability to absorb carbon from the atmosphere on a regional scale and reducing oceanic biodiversity globally.

Arctic methane emissions indirectly caused by anthropogenic global warming also affect the carbon cycle, and contribute to further warming in what is known as climate change feedback.

On 12 November 2015, NASA scientists reported that human-made carbon dioxide (CO_2) continues to increase above levels not seen in hundreds of thousands of years: currently, about half of the carbon dioxide released from the burning of fossil fuels remains in the atmosphere and is not absorbed by vegetation and the oceans.

Phosphorus Cycle

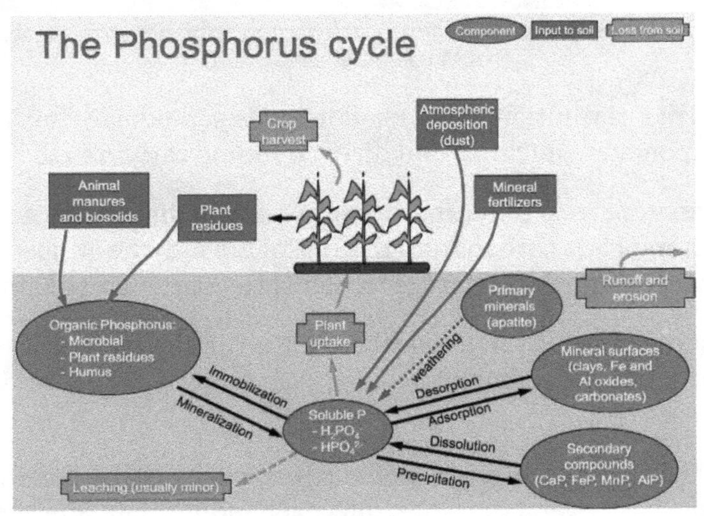

Phosphorus Cycle on land

The phosphorus cycle is the biogeochemical cycle that describes the movement of phosphorus through the lithosphere, hydrosphere, and biosphere. Unlike many other biogeochemical cycles, the atmosphere does not play a significant role in the movement of phosphorus, because phosphorus and phosphorus-based compounds are usually solids at the typical ranges of temperature and pressure found on Earth. The production of phosphine gas occurs in only specialized, local conditions.

On the land, phosphorus (chemical symbol, P) gradually becomes less available to plants over thousands of years, because it is slowly lost in runoff. Low concentration of P in soils reduces plant growth, and slows soil microbial growth - as shown in studies of soil microbial biomass. Soil microorganisms act as both sinks and sources of available P in the biogeochemical cycle. Locally, transformations of P are chemical, biological and microbiological: the major long-term transfers in the global cycle, however, are driven by tectonic movements in geologic time.

Humans have caused major changes to the global P cycle through shipping of P minerals, and use of P fertilizer, and also the shipping of food from farms to cities, where it is lost as effluent

Phosphorus in the Environment

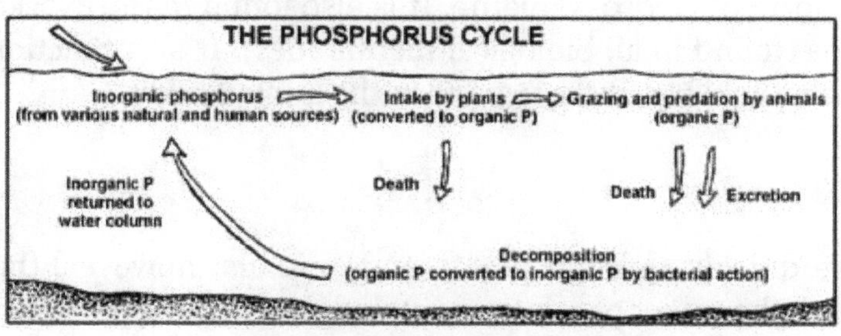

The aquatic phosphorus cycle

Ecological Function

Phosphorus is an essential nutrient for plants and animals. Phosphorus is a limiting nutrient for aquatic organisms. Phosphorus forms parts of important life-sustaining molecules that are very common in the biosphere. Phosphorus does not enter the atmosphere, remaining mostly on land and in rock and soil minerals. Eighty percent of the mined phosphorus is used to make fertilizers. Phosphates from fertilizers, sewage and detergents can cause pollution in lakes and streams. Overenrichment of phosphate in both fresh and inshore marine waters can lead to massive algae blooms which, when they die and decay, leads to eutrophication of fresh waters only. An example of this is the Canadian Experimental Lakes Area. These freshwater algal blooms should not be confused with those in saltwater environments. Recent research suggests that the predominant pollutant responsible for algal blooms in salt water estuaries and coastal marine habitats is Nitrogen.

Phosphorus occurs most abundantly in nature as part of the orthophosphate ion $(PO_4)^{3-}$, consisting of a P atom and 4 oxygen atoms. On land most phosphorus is found in rocks and minerals. Phosphorus rich deposits have generally formed in the ocean or from guano, and over time, geologic processes bring ocean sediments to land. Weathering of rocks and minerals release phosphorus in a soluble form where it is taken up by plants, and it is transformed into organic compounds. The plants may then be consumed by herbivores and the phosphorus is either incorporated into their tissues or excreted. After death, the animal or plant decays, and phosphorus is returned to the soil where a large part of the phosphorus is transformed into insoluble compounds. Runoff may carry a small part of the phosphorus back to the ocean. Generally with time (thousands of years) soils become deficient in phosphorus leading to ecosystem retrogression.

Biological Function

The primary biological importance of phosphates is as a component of nucleotides, which serve as energy storage within cells (ATP) or when linked together, form the nucleic acids DNA and RNA. The double helix of our DNA is only possible because of the phosphate ester bridge that binds the helix. Besides making biomolecules, phosphorus is also found in bone and the enamel of mammalian teeth, whose strength is derived from calcium phosphate in the form of Hydroxylapatite. It is also found in the exoskeleton of insects, and phospholipids (found in all biological membranes). It also functions as a buffering agent in maintaining acid base homeostasis in the human body.

Process of the Cycle

Phosphates move quickly through plants and animals; however, the processes that move them through the soil or ocean are very slow, making the phosphorus cycle overall one of the slowest biogeochemical cycles.

Initially, phosphate weathers from rocks and minerals, the most common mineral being apatite. Overall small losses occur in terrestrial environments by leaching and erosion, through the action of rain. In soil, phosphate is absorbed on iron oxides, aluminium hydroxides, clay surfaces, and organic matter particles, and becomes incorporated (immobilized or fixed). Plants and fungi can also be active in making P soluble.

Unlike other cycles, P is not normally found in the air as a gas; it only occurs under highly reducing conditions as the gas phosphine PH_3.

Phosphatic Minerals

The availability of phosphorus in an ecosystem is restricted by the rate of release of this element during weathering. The release of phosphorus from apatite dissolution is a key control on ecosystem productivity. The primary mineral with significant phosphorus content, apatite [$Ca_5(PO_4)_3OH$] undergoes carbonation.

Little of this released phosphorus is taken up by biota (organic form), whereas a larger proportion reacts with other soil minerals. This leads to precipitation into unavailable forms in the later stage of weathering and soil development. Available phosphorus is found in a biogeochemical cycle in the upper soil profile, while phosphorus found at lower depths is primarily involved in geochemical reactions with secondary minerals. Plant growth depends on the rapid root uptake of phosphorus released from dead organic matter in the biochemical cycle. Phosphorus is limited in supply for plant growth. Phosphates move quickly through plants and animals; however, the processes that move them through the soil or ocean are very slow, making the phosphorus cycle overall one of the slowest biogeochemical cycles.

Low-molecular-weight (LMW) organic acids are found in soils. They originate from the activities of various microorganisms in soils or may be exuded from the roots of living plants. Several of those organic acids are capable of forming stable organo-metal complexes with various metal ions found in soil solutions. As a result, these processes may lead to the release of inorganic phosphorus associated with aluminium, iron, and calcium in soil minerals. The production and release of oxalic acid by mycorrhizal fungi explain their importance in maintaining and supplying phosphorus to plants.

The availability of organic phosphorus to support microbial, plant and animal growth depends on the rate of their degradation to generate free phosphate. There are various enzymes such as phosphatases, nucleases and phytase involved for the degradation. Some of the abiotic pathways in the environment studied are hydrolytic reactions and photolytic reactions. Enzymatic hydrolysis of organic phosphorus is an essential step in the biogeochemical phosphorus cycle, including the phosphorus nutrition of plants and microorganisms and the transfer of organic phosphorus from soil to bodies of water. Many organisms rely on the soil derived phosphorus for their phosphorus nutrition.

Human Influences

Nutrients are important to the growth and survival of living organisms, and hence, are essential for development and maintenance of healthy ecosystems. Humans have greatly influenced the phosphorus cycle by mining phosphorus, converting it to fertilizer, and by shipping fertilizer and products around the globe. Transporting phosphorus in food from farms to cities has made a major change in the global Phosphorus cycle. However, excessive amounts of nutrients, particularly phosphorus and nitrogen, are detrimental to aquatic ecosystems. Waters are enriched in phosphorus from farms' run-off, and from effluent that is inadequately treated before it is discharged to waters. Natural eutrophication is a process by which lakes gradually age and become more productive and may take thousands of years to progress. Cultural or anthropogenic eutrophication, however, is water pollution caused by excessive plant nutrients; this results in excessive growth in the algal population; when this algae dies its putrefaction depletes the water of oxygen. Such eutrophication may also give rise to toxic algal bloom.Both these effects cause animal and plant death rates to increase as the plants take in poisonous water while the animals drink the poisoned water. Surface and subsurface runoff and erosion from high-phosphorus soils may be major contributing factors to this fresh water eutrophication. The processes controlling soil Phosphorus release to surface runoff and to subsurface flow are a complex interaction between the type of phosphorus input, soil type and management, and transport processes depending on hydrological conditions.

Repeated application of liquid hog manure in excess to crop needs can have detrimental effects on soil phosphorus status. Also, application of biosolids may increase available phosphorus in soil. In poorly drained soils or in areas where snowmelt can cause periodic waterlogging, dereducing conditions can be attained in 7–10 days. This causes a sharp increase in phosphorus concentration in solution and phosphorus can be leached. In addition, reduction of the soil causes a shift in phosphorus from resilient to more labile forms. This could eventually increase the potential for phosphorus loss. This is of particular concern for the environmentally sound management of such areas, where disposal of agricultural wastes has already become a problem. It is suggested that the water regime of soils that are to be used for organic wastes disposal is taken into account in the preparation of waste management regulations.

Human interference in the phosphorus cycle occurs by overuse or careless use of phosphorus fertilizers. This results in increased amounts of phosphorus as pollutants in bodies of water resulting in eutrophication. Eutrophication devastates water ecosystems by inducing anoxic conditions.

Water Cycle

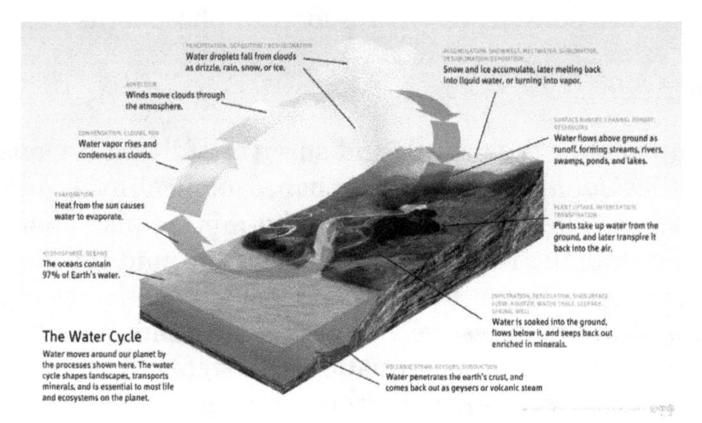

Diagram of the Water Cycle

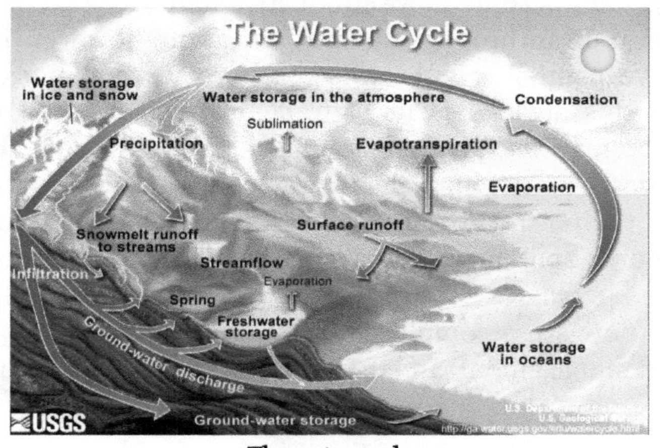

The water cycle

The water cycle, also known as the hydrological cycle or the H_2O cycle, describes the continuous movement of water on, above and below the surface of the Earth. The mass of water on Earth remains fairly constant over time but the partitioning of the water into the major reservoirs of ice, fresh water, saline water and atmospheric water is variable depending on a wide range of climatic variables. The water moves from one reservoir to another, such as from river to ocean, or from the ocean to the atmosphere, by the physical processes of evaporation, condensation, precipitation, infiltration, surface runoff, and subsurface flow. In doing so, the water goes through different phases: liquid, solid (ice) and vapor.

The water cycle involves the exchange of energy, which leads to temperature changes. For instance, when water evaporates, it takes up energy from its surroundings and cools the environment. When it condenses, it releases energy and warms the environment. These heat exchanges influence climate.

The evaporative phase of the cycle purifies water which then replenishes the land with freshwater. The flow of liquid water and ice transports minerals across the globe. It is also involved in reshaping the geological features of the Earth, through processes including erosion and sedimentation. The water cycle is also essential for the maintenance of most life and ecosystems on the planet.

Description

The sun, which drives the water cycle, heats water in oceans and seas. Water evaporates as water vapor into the air. Ice and snow can sublimate directly into water vapour. Evapotranspiration is water transpired from plants and evaporated from the soil. The water vapour molecule H_2O, has less density compared to the major components of the atmosphere, nitrogen and oxygen, N_2 and O_2. Due to the significant difference in molecular mass, water vapor in gas form gain height in open air as a result of buoyancy. However, as altitude increases, air pressure decreases and the temperature drops. The lowered temperature causes water vapour to condense into a tiny liquid water droplet which is heavier than the air, such that it falls unless supported by an updraft. A huge concentration of these droplets over a large space up in the atmosphere become visible as cloud. Fog is formed if the water vapour condense near ground level, as a result of moist air and cool air collision or an abrupt reduction in air pressure. Air currents move water vapour around the globe, cloud particles collide, grow, and fall out of the upper atmospheric layers as precipitation. Some precipitation falls as snow or hail, sleet, and can accumulate as ice caps and glaciers, which can store frozen water for thousands of years. Most water falls back into the oceans or onto land as rain, where the water flows over the ground as surface runoff. A portion of runoff enters rivers in valleys in the landscape, with streamflow moving water towards the oceans. Runoff and water emerging from the ground (groundwater) may be stored as freshwater in lakes. Not all runoff flows into rivers, much of it soaks into the ground as infiltration. Some water infiltrates deep into the ground and replenishes aquifers, which can store freshwater for long periods of time. Some infiltration stays close to the land surface and can seep back into surface-water bodies (and the ocean) as groundwater discharge. Some groundwater finds openings in the land surface and comes out as freshwater springs. In river valleys and flood-plains there is often continuous water exchange between surface water and ground water in the hyporheic zone. Over time, the water returns to the ocean, to continue the water cycle.

Processes

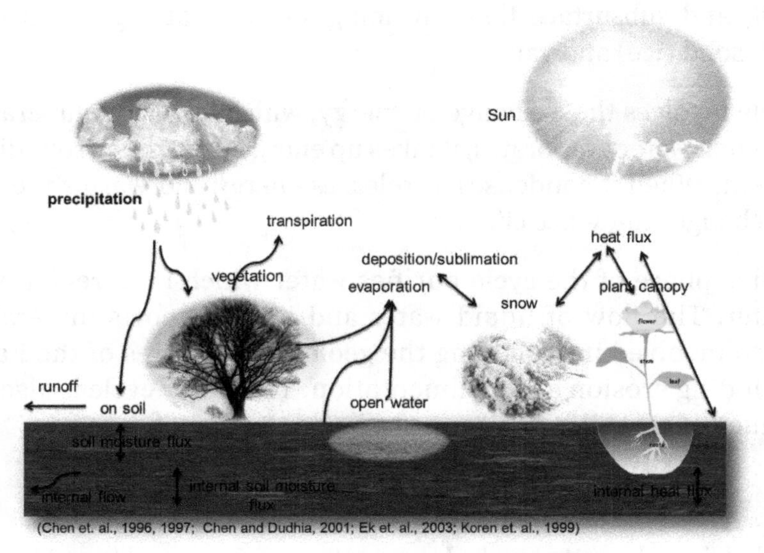

Many different processes lead to movements and phase changes in water

Precipitation

Condensed water vapor that falls to the Earth's surface. Most precipitation occurs as rain, but also includes snow, hail, fog drip, graupel, and sleet. Approximately 505,000 km³ (121,000 cu mi) of water falls as precipitation each year, 398,000 km³ (95,000 cu mi) of it over the oceans. The rain on land contains 107,000 km³ (26,000 cu mi) of water per year and a snowing only 1,000 km³ (240 cu mi). 78% of global precipitation occurs over the ocean.

Canopy interception

The precipitation that is intercepted by plant foliage, eventually evaporates back to the atmosphere rather than falling to the ground.

Snowmelt

The runoff produced by melting snow.

Runoff

The variety of ways by which water moves across the land. This includes both surface runoff and channel runoff. As it flows, the water may seep into the ground, evaporate into the air, become stored in lakes or reservoirs, or be extracted for agricultural or other human uses.

Infiltration

The flow of water from the ground surface into the ground. Once infiltrated, the water becomes soil moisture or groundwater. A recent global study using water stable isotopes, however, shows that not all soil moisture is equally available for groundwater recharge or for plant transpiration.

Subsurface flow

The flow of water underground, in the vadose zone and aquifers. Subsurface water may return to the surface (e.g. as a spring or by being pumped) or eventually seep into the oceans. Water returns to the land surface at lower elevation than where it infiltrated, under the force of gravity or gravity induced pressures. Groundwater tends to move slowly, and is replenished slowly, so it can remain in aquifers for thousands of years.

Evaporation

The transformation of water from liquid to gas phases as it moves from the ground or bodies of water into the overlying atmosphere. The source of energy for evaporation is primarily solar radiation. Evaporation often implicitly includes transpiration from plants, though together they are specifically referred to as evapotranspiration. Total annual evapotranspiration amounts to approximately 505,000 km^3 (121,000 cu mi) of water, 434,000 km^3 (104,000 cu mi) of which evaporates from the oceans. 86% of global evaporation occurs over the ocean.

Sublimation

The state change directly from solid water (snow or ice) to water vapor.

Deposition

This refers to changing of water vapor directly to ice.

Advection

The movement of water — in solid, liquid, or vapor states — through the atmosphere. Without advection, water that evaporated over the oceans could not precipitate over land.

Condensation

The transformation of water vapor to liquid water droplets in the air, creating clouds and fog.

Transpiration

The release of water vapor from plants and soil into the air. Water vapor is a gas that cannot be seen.

Percolation

Water flows vertically through the soil and rocks under the influence of gravity

Plate tectonics

Water enters the mantle via subduction of oceanic crust. Water returns to the surface via volcanism.

Water cycle thus involves many of the intermediate processes.

Residence Times

Average reservoir residence times	
Reservoir	Average residence time
Antarctica	20,000 years
Oceans	3,200 years
Glaciers	20 to 100 years
Seasonal snow cover	2 to 6 months
Soil moisture	1 to 2 months
Groundwater: shallow	100 to 200 years
Groundwater: deep	10,000 years
Lakes	50 to 100 years
Rivers	2 to 6 months
Atmosphere	9 days

The *residence time* of a reservoir within the hydrologic cycle is the average time a water molecule will spend in that reservoir. It is a measure of the average age of the water in that reservoir.

Groundwater can spend over 10,000 years beneath Earth's surface before leaving. Particularly old groundwater is called fossil water. Water stored in the soil remains there very briefly, because it is spread thinly across the Earth, and is readily lost by evaporation, transpiration, stream flow, or groundwater recharge. After evaporating, the residence time in the atmosphere is about 9 days before condensing and falling to the Earth as precipitation.

The major ice sheets - Antarctica and Greenland - store ice for very long periods. Ice from Antarctica has been reliably dated to 800,000 years before present, though the average residence time is shorter.

In hydrology, residence times can be estimated in two ways. The more common method relies on the principle of conservation of mass and assumes the amount of water in a given reservoir is roughly constant. With this method, residence times are estimated by dividing the volume of the reservoir by the rate by which water either enters or exits the reservoir. Conceptually, this is equivalent to timing how long it would take the reservoir to become filled from empty if no water were to leave (or how long it would take the reservoir to empty from full if no water were to enter).

An alternative method to estimate residence times, which is gaining in popularity for dating groundwater, is the use of isotopic techniques. This is done in the subfield of isotope hydrology.

Changes Over Time

The water cycle describes the processes that drive the movement of water throughout the hydrosphere. However, much more water is "in storage" for long periods of time than is actually moving through the cycle. The storehouses for the vast majority of all water on Earth are the oceans. It is estimated that of the 332,500,000 mi³ (1,386,000,000 km³) of the world's water supply, about 321,000,000 mi³ (1,338,000,000 km³) is stored in oceans, or about 97%. It is also estimated that the oceans supply about 90% of the evaporated water that goes into the water cycle.

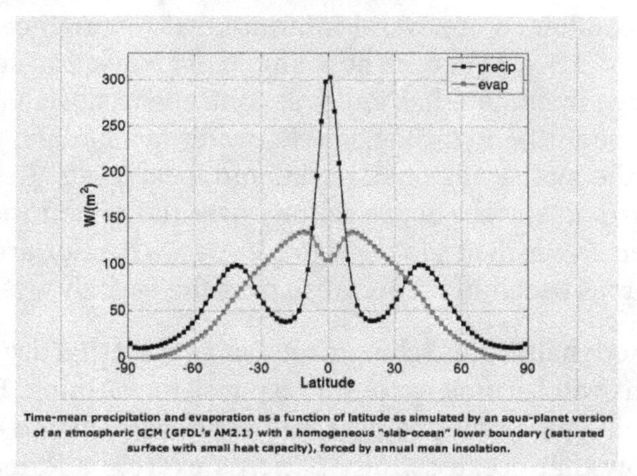

Time-mean precipitation and evaporation as a function of latitude as simulated by an aqua-planet version of an atmospheric GCM (GFDL's AM2.1) with a homogeneous "slab-ocean" lower boundary (saturated surface with small heat capacity), forced by annual mean insolation.

Time-mean precipitation and evaporation as a function of latitude as simulated by an aqua-planet version of an atmospheric GCM (GFDL's AM2.1) with a homogeneous "slab-ocean" lower boundary (saturated surface with small heat capacity), forced by annual mean insolation.

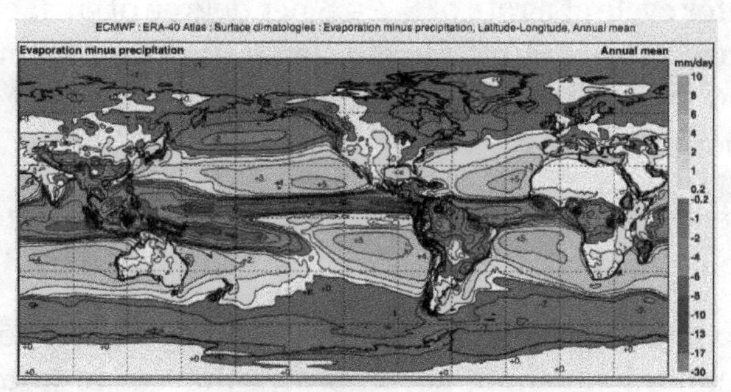

Global map of annual mean evaporation minus precipitation by latitude-longitude

During colder climatic periods more ice caps and glaciers form, and enough of the global water supply accumulates as ice to lessen the amounts in other parts of the water cycle. The reverse is true during warm periods. During the last ice age glaciers covered almost one-third of Earth's land mass, with the result being that the oceans were about 400 ft (122 m) lower than today. During the last global "warm spell," about 125,000 years ago, the seas were about 18 ft (5.5 m) higher than they are now. About three million years ago the oceans could have been up to 165 ft (50 m) higher.

The scientific consensus expressed in the 2007 Intergovernmental Panel on Climate Change (IPCC) Summary for Policymakers is for the water cycle to continue to intensify throughout the 21st century, though this does not mean that precipitation will increase in all regions. In subtropical land areas — places that are already relatively dry — precipitation is projected to decrease during the 21st century, increasing the probability of drought. The drying is projected to be strongest near the poleward margins of the subtropics (for example, the Mediterranean Basin, South Africa, southern Australia, and the Southwestern United States). Annual precipitation amounts are expected to increase in near-equatorial regions that tend to be wet in the present climate, and also at high latitudes. These large-scale patterns are present in nearly all of the climate model simulations conducted at several international research centers as part of the 4th Assessment of the IPCC. There is now ample evidence that increased hydrologic variability and change in climate has and will continue to have a profound impact on the water sector through the hydrologic cycle, water availability, water demand, and water allocation at the global, regional, basin, and local levels. Research published in 2012 in *Science* based on surface ocean salinity over the period 1950 to 2000 confirm this projection of an intensified global water cycle with salty areas becoming more saline and fresher areas becoming more fresh over the period:

Fundamental thermodynamics and climate models suggest that dry regions will become drier and wet regions will become wetter in response to warming. Efforts to detect this long-term response in sparse surface observations of rainfall and evaporation remain ambiguous. We show that ocean salinity patterns express an identifiable fingerprint of an intensifying water cycle. Our 50-year observed global surface salinity changes, combined with changes from global climate models, present robust evidence of an intensified global water cycle at a rate of 8 ± 5% per degree of surface warming. This rate is double the response projected by current-generation climate models and suggests that a substantial (16 to 24%) intensification of the global water cycle will occur in a future 2° to 3° warmer world.

An instrument carried by the SAC-D satellite launched in June, 2011 measures global sea surface salinity but data collection began only in June, 2011.

Glacial retreat is also an example of a changing water cycle, where the supply of water to glaciers from precipitation cannot keep up with the loss of water from melting and sublimation. Glacial retreat since 1850 has been extensive.

Human activities that alter the water cycle include:

- agriculture
- industry
- alteration of the chemical composition of the atmosphere
- construction of dams
- deforestation and afforestation
- removal of groundwater from wells
- water abstraction from rivers
- urbanization

Effects on Climate

The water cycle is powered from solar energy. 86% of the global evaporation occurs from the oceans, reducing their temperature by evaporative cooling. Without the cooling, the effect of evaporation on the greenhouse effect would lead to a much higher surface temperature of 67 °C (153 °F), and a warmer planet.

Aquifer drawdown or overdrafting and the pumping of fossil water increases the total amount of water in the hydrosphere, and has been postulated to be a contributor to sea-level rise.

Effects on Biogeochemical Cycling

While the water cycle is itself a biogeochemical cycle, flow of water over and beneath the Earth is a key component of the cycling of other biogeochemicals. Runoff is responsible for almost all of the transport of eroded sediment and phosphorus from land to waterbodies. The salinity of the oceans is derived from erosion and transport of dissolved salts from the land. Cultural eutrophication of lakes is primarily due to phosphorus, applied in excess to agricultural fields in fertilizers, and then transported overland and down rivers. Both runoff and groundwater flow play significant roles in transporting nitrogen from the land to waterbodies. The dead zone at the outlet of the Mississippi River is a consequence of nitrates from fertilizer being carried off agricultural fields and funnelled down the river system to the Gulf of Mexico. Runoff also plays a part in the carbon cycle, again through the transport of eroded rock and soil.

Slow Loss Over Geologic Time

The hydrodynamic wind within the upper portion of a planet's atmosphere allows light chemical elements such as Hydrogen to move up to the exobase, the lower limit of the exosphere, where the gases can then reach escape velocity, entering outer space without impacting other particles of gas. This type of gas loss from a planet into space is known as planetary wind. Planets with hot lower atmospheres could result in humid upper atmospheres that accelerate the loss of hydrogen.

History of Hydrologic Cycle Theory

Floating Land Mass

In ancient times, it was thought that the land mass floated on a body of water, and that most of the water in rivers has its origin under the earth. Examples of this belief can be found in the works of Homer (circa 800 BCE).

Source of Rain

In the ancient near east, Hebrew scholars observed that even though the rivers ran into the sea, the sea never became full (Ecclesiastes 1:7). Some scholars conclude that the water cycle was described completely during this time in this passage: "The wind goeth toward the south, and turneth about unto the north; it whirleth about continually, and the wind returneth again according to its circuits. All the rivers run into the sea; yet the sea is not full; unto the place from whence the rivers come, thither they return again" (Ecclesiastes 1:6-7, KJV). Scholars are not in agreement as to the date of Ecclesiastes, though most scholars point to a date during the time of Solomon, the son of David and Bathsheba, "three thousand years ago, there is some agreement that the time period is 962-922 BCE. Furthermore, it was also observed that when the clouds were full, they emptied rain on the earth (Ecclesiastes 11:3). In addition, during 793-740 BC a Hebrew prophet, Amos, stated that water comes from the sea and is poured out on the earth (Amos 5:8, 9:6).

Precipitation and Percolation

In the Adityahridayam (a devotional hymn to the Sun God) of Ramayana, a Hindu epic dated to the 4th century BC, it is mentioned in the 22nd verse that the Sun heats up water and sends it down as rain. By roughly 500 BCE, Greek scholars were speculating that much of the water in rivers can be attributed to rain. The origin of rain was also known by then. These scholars maintained the belief, however, that water rising up through the earth contributed a great deal to rivers. Examples of this thinking included Anaximander (570 BCE) (who also speculated about the evolution of land animals from fish) and Xenophanes of Colophon (530 BCE). Chinese scholars such as Chi Ni Tzu (320 BC) and Lu Shih Ch'un Ch'iu (239 BCE) had similar thoughts. The idea that the water cycle is a closed cycle can be found in the works of Anaxagoras of Clazomenae (460 BCE) and Diogenes of Apollonia (460 BCE). Both Plato (390 BCE) and Aristotle (350 BCE) speculated about percolation as part of the water cycle.

Precipitation Alone

In the Biblical Book of Job, dated between 7th and 2nd centuries BCE, there is a description of precipitation in the hydrologic cycle, "For he maketh small the drops of water: they pour down rain according to the vapour thereof; Which the clouds do drop and distil upon man abundantly" (Job 36:27-28, KJV). Also found in the book of Ecclesiastes "All the rivers flow into the sea, Yet the sea is not full. To the place where the rivers flow, There they flow again."

Up to the time of the Renaissance, it was thought that precipitation alone was insufficient to feed rivers, for a complete water cycle, and that underground water pushing upwards from the oceans were the main contributors to river water. Bartholomew of England held this view (1240 CE), as did Leonardo da Vinci (1500 CE) and Athanasius Kircher (1644 CE).

The first published thinker to assert that rainfall alone was sufficient for the maintenance of rivers was Bernard Palissy (1580 CE), who is often credited as the "discoverer" of the modern theory of the water cycle. Palissy's theories were not tested scientifically until 1674, in a study commonly attributed to Pierre Perrault. Even then, these beliefs were not accepted in mainstream science until the early nineteenth century.

Biological Pump

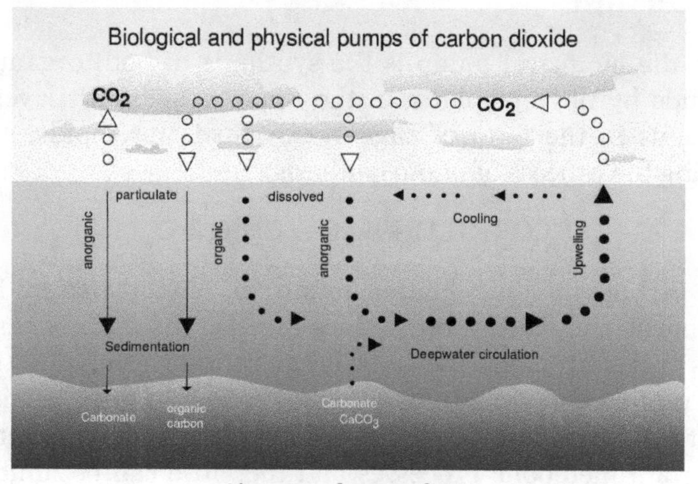

Air-sea exchange of CO_2

The biological pump, in its simplest form, is the ocean's biologically driven sequestration of carbon from the atmosphere to the deep sea. It is the part of the oceanic carbon cycle responsible for the cycling of organic matter formed by phytoplankton during photosynthesis (soft-tissue pump), as well as the cycling of calcium carbonate (CaCO3) formed by certain plankton and mollusks as a protective coating (carbonate pump).

Overview

The biological pump can be divided into three distinct phases, the first of which is the production of fixed carbon by planktonic phototrophs in the euphotic (sunlit) surface region of the ocean. In these surface waters, phytoplankton use carbon dioxide (CO_2), nitrogen (N), phosphorus (P), and other trace elements (barium, iron, zinc, etc.) during photosynthesis to make carbohydrates, lipids, and proteins. Some plankton, (e.g. coccolithophores and foraminifera) combine calcium (Ca) and dissolved carbonates (carbonic acid and bicarbonate) to form a calcium carbonate ($CaCO_3$) protective coating.

Once this carbon is fixed into soft or hard tissue, the organisms either stay in the euphotic zone to be recycled as part of the regenerative nutrient cycle or once they die, continue to the second phase of the biological pump and begin to sink to the ocean floor. The sinking particles will often form aggregates as they sink, greatly increasing the sinking rate. It is this aggregation that gives particles a better chance of escaping predation and decomposition in the water column and eventually make it to the sea floor.

The fixed carbon that is either decomposed by bacteria on the way down or once on the sea floor then enters the final phase of the pump and is remineralized to be used again in primary production. The particles that escape these processes entirely are sequestered in the sediment and may remain there for thousands of years. It is this sequestered carbon that is responsible for ultimately lowering atmospheric CO_2.

Primary Production

The first step in the biological pump is the synthesis of both organic and inorganic carbon compounds by phytoplankton in the uppermost, sunlit layers of the ocean. Organic compounds in the form of sugars, carbohydrates, lipids, and proteins are synthesized during the process of photosynthesis:

$$CO_2 + H_2O + \text{light} \rightarrow CH_2O + O_2$$

In addition to carbon, organic matter found in phytoplankton is composed of nitrogen, phosphorus and various other trace metals. The ratio of carbon to nitrogen and phosphorus varies little and has an average ratio of 106C:16N:1P, known as the Redfield ratio. Trace metals such as magnesium, cadmium, iron, calcium, barium and copper are orders of magnitude less prevalent in phytoplankton organic material, but necessary for certain metabolic processes and therefore can be limiting nutrients in photosynthesis due to their lower abundance in the water column.

Oceanic primary production accounts for about half of the carbon fixation carried out on Earth. Approximately 50-60 Pg of carbon are fixed by marine phytoplankton each year despite the fact that they comprise less than 1% of the total photosynthetic biomass on Earth. The majority of this carbon fixation (~80%) is carried out in the open ocean while the remaining amount occurs in the very productive upwelling regions of the ocean. Despite these productive regions producing 2 to 3 times as much fixed carbon per area, the open ocean accounts for greater than 90% of the ocean area and therefore is the larger contributor.

Calcium Carbonate

Carbon is also biologically fixed in the form of calcium carbonate ($CaCO_3$) used as a protective coating for many planktonic species (coccolithophores, foraminifera) as well as larger marine organisms (mollusk shells). While this form of carbon is not directly taken from the atmospheric budget, it is formed from dissolved forms of carbonate which are in equilibrium with CO_2 and then responsible for removing this carbon via sequestration.

$$CO_2 + H_2O \rightarrow H_2CO_3 \rightarrow H^+ + HCO_3^-$$

$$Ca^{2+} + 2HCO_3^- \rightarrow CaCO_3 + CO_2 + H_2O$$

While this process does manage to fix a large amount of carbon, two units of alkalinity are sequestered for every unit of sequestered carbon, thereby lowering the pH of surface water and raising atmospheric CO_2. The formation and sinking of $CaCO_3$ drives a surface to deep alkalinity gradient which serves to raise the partial pressure of dissolved CO_2 in surface waters and actually raise atmospheric levels. In addition, the sequestration of $CaCO_3$ serves to lower overall oceanic alkalinity and again raise atmospheric levels.

Marine Snow

The vast majority of carbon incorporated in organic and inorganic biological matter is formed at the sea surface and then must sink to the ocean floor. A single phytoplankton cell has a sinking rate around 1 m per day and with 4000 m as the average depth of the ocean, it can take over ten years for these cells to reach the ocean floor. However, through processes such as coagulation and expulsion in predator fecal pellets, these cells form aggregates. These aggregates, known as marine snow, have sinking rates orders of magnitude greater than individual cells and complete their journey to the deep in a matter of days.

White Cliffs of Dover

Of the 50-60 Pg of carbon fixed annually, roughly 10% leaves the surface mixed layer of the oceans, while less than 0.5% of eventually reaches the sea floor. Most is retained in regenerated production in the euphotic zone and a significant portion is remineralized in midwater processes during particle sinking. The portion of carbon that leaves the surface mixed layer of the ocean is sometimes considered "sequestered", and essentially removed from contact with the atmosphere for many centuries. However, work also finds that, in regions such as the Southern Ocean, much of this carbon can quickly (within decades) come back into contact with the atmosphere. The portion of carbon that makes it to the sea floor becomes part of the geologic record and in the case of the calcium carbonate, may form large deposits and resurface through tectonic motion as in the case with the White Cliffs of Dover in Southern England. These cliffs are made almost entirely of the plates of buried coccolithophores.

Quantification

As the biological pump plays an important role in the Earth's carbon cycle, significant effort is spent quantifying its strength. However, because they occur as a result of poorly constrained ecological interactions usually at depth, the processes that form the biological pump are difficult to measure. A common method is to estimate primary production fuelled by nitrate and ammonium as these nutrients have different sources that are related to the remineralisation of sinking material. From these it is possible to derive the so-called f-ratio, a proxy for the local strength of the biological pump. Applying the results of local studies to the global scale are complicated by the role the ocean's circulation plays in different ocean regions.

The biological pump has a physico-chemical counterpart known as the solubility pump. For an overview of both pumps.

Anthropogenic Changes

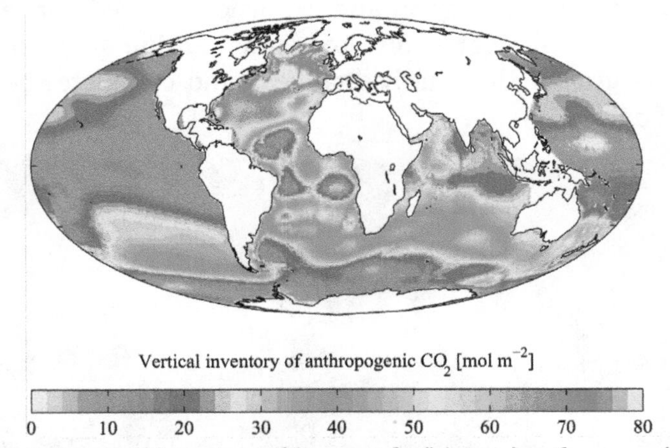

Estimated vertical inventory of "present day" (1990s) anthropogenic CO_2

It was recently determined that coccolithophore concentrations in the North Atlantic have increased by an order of magnitude since the 1960s and an increase in absorbed CO2, as well as temperature, were modeled to be the most likely cause of this increase.

Changes in land use, the combustion of fossil fuels, and the production of cement have led to an increase in CO_2 concentration in the atmosphere. At present, about one third (approximately 2 Pg C y^{-1} = 2 × 10^{15} grams of carbon per year) of anthropogenic emissions of CO_2 are believed to be entering the ocean. However, the biological pump is not believed to play a significant role in the net uptake of CO_2 by oceans. This is because the biological pump is primarily limited by the availability of light and nutrients, and not by carbon. This is in contrast to the situation on land, where elevated atmospheric concentrations of CO_2 may increase primary production because land plants are able to improve their water-use efficiency (= decrease transpiration) when CO_2 is easier to obtain. However, there are still considerable uncertainties in the marine carbon cycle, and some research suggests that a link between elevated CO_2 and marine primary production exists.

However, climate change may affect the biological pump in the future by warming and stratifying the surface ocean. It is believed that this could decrease the supply of nutrients to the euphotic zone, reducing primary production there. Also, changes in the ecological success of calcifying organisms caused by ocean acidification may affect the biological pump by altering the strength of the hard tissues pump. This may then have a "knock-on" effect on the soft tissues pump because calcium carbonate acts to ballast sinking organic material.

Carbon Sequestration

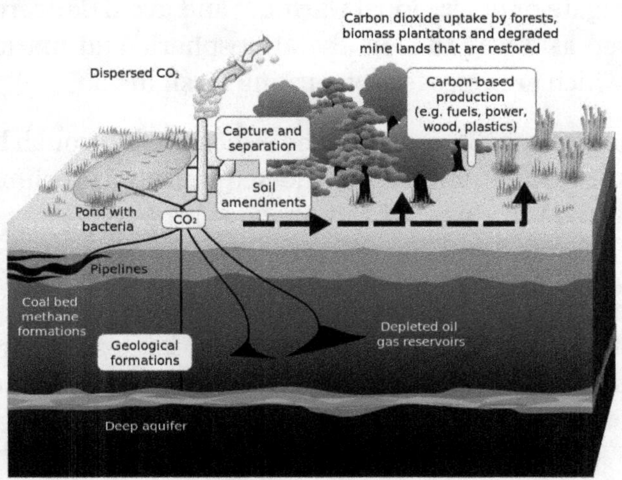

Schematic showing both terrestrial and geological sequestration of carbon dioxide emissions from a coal-fired plant.

Carbon sequestration is the process involved in carbon capture and the long-term storage of atmospheric carbon dioxide. Carbon sequestration involves long-term storage of carbon dioxide or other forms of carbon to either mitigate or defer global warming and avoid dangerous climate change. It has been proposed as a way to slow the atmospheric and marine accumulation of greenhouse gases, which are released by burning fossil fuels.

Carbon dioxide (CO_2) is naturally captured from the atmosphere through biological, chemical, and physical processes. Artificial processes have been devised to produce similar effects, including large-scale, artificial capture and sequestration of industrially produced CO_2 using subsurface saline aquifers, reservoirs, ocean water, aging oil fields, or other carbon sinks.

Description

Carbon sequestration is the process involved in carbon capture and the long-term storage of atmospheric carbon dioxide (CO_2) and may refer specifically to:

- "The process of removing carbon from the atmosphere and depositing it in a reservoir." When carried out deliberately, this may also be referred to as carbon dioxide removal, which is a form of geoengineering.
- Carbon capture and storage, where carbon dioxide is removed from flue gases (e.g., at power stations) before being stored in underground reservoirs.
- Natural biogeochemical cycling of carbon between the atmosphere and reservoirs, such as by chemical weathering of rocks.

Carbon sequestration describes long-term storage of carbon dioxide or other forms of carbon to either mitigate or defer global warming and avoid dangerous climate change. It has been proposed as a way to slow the atmospheric and marine accumulation of greenhouse gases, which are released by burning fossil fuels.

Carbon dioxide is naturally captured from the atmosphere through biological, chemical or physical processes. Some anthropogenic sequestration techniques exploit these natural processes, while some use entirely artificial processes.

Carbon dioxide may be captured as a pure by-product in processes related to petroleum refining or from flue gases from power generation. CO_2 sequestration includes the storage part of carbon capture and storage, which refers to large-scale, artificial capture and sequestration of industrially produced CO_2 using subsurface saline aquifers, reservoirs, ocean water, aging oil fields, or other carbon sinks.

Biological Processes

Biosequestration or carbon sequestration through biological processes affects the global carbon cycle. Examples include major climatic fluctuations, such as the Azolla event, which created the current Arctic climate. Such processes created fossil fuels, as well as clathrate and limestone. By manipulating such processes, geoengineers seek to enhance sequestration.

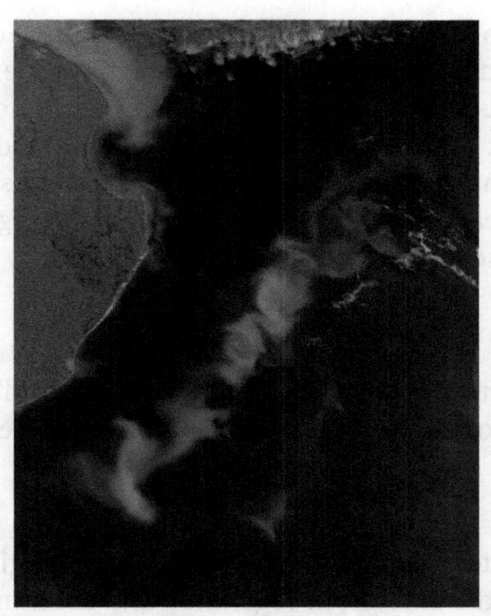

An oceanic phytoplankton bloom in the South Atlantic Ocean, off the coast of Argentina. Encouraging such blooms with iron fertilization could lock up carbon on the seabed.

Peat Production

Peat bogs are a very important carbon store. Peat bogs act as a sink for carbon due to the accumulation of partially decayed biomass that would otherwise continue to decay completely. There is a variance on how much the peatlands act as a carbon sink or carbon source that can be linked to varying climates in different areas of the world and different times of the year. By creating new bogs, or enhancing existing ones, the amount of carbon that is sequestered by bogs would increase.

Forestry

Reforestation is the replanting of trees on marginal crop and pasture lands to incorporate carbon from atmospheric CO2 into biomass. For this process to succeed the carbon must not return to the atmosphere from mass burning or rotting when the trees die. To this end, land allotted to the trees must not be converted to other uses and management of the frequency of disturbances might be necessary in order to avoid extreme events. Alternatively, the wood from them must itself be sequestered, e.g., via biochar, bio-energy with carbon storage (BECS), landfill or 'stored' by use in e.g. construction. Short of growth in perpetuity, however, reforestation with long-lived trees (>100 years) will sequester carbon for a more graduated release, minimizing impact during the expected carbon crisis of the 21st century.

Urban Forestry

Urban Forestry increases the amount of carbon taken up in cities by adding new tree sites and the sequestration of carbon occurs over the lifetime of the tree. It is generally practiced and maintained on smaller scales, like in cities. The results of urban forestry can have different results depending on the type of vegetation that is being used, so it can function as a sink but can also function as a source of emissions. Along with sequestration by the plants which is difficult to measure but seems to have little effect on the overall amount of carbon dioxide that is uptaken, the vegetation can have indirect effects on carbon by reducing need for energy consumption.

Wetland Restoration

Wetland soil is an important carbon sink; 14.5% of the world's soil carbon is found in wetlands, while only 6% of the world's land is composed of wetlands.

Agriculture

Globally, soils are estimated to contain approximately 1,500 gigatons of organic carbon to 1 m depth, more than the amount in vegetation and the atmosphere.

Modification of agricultural practices is a recognized method of carbon sequestration as soil can act as an effective carbon sink offsetting as much as 20% of 2010 carbon dioxide emissions annually.

Carbon emission reduction methods in agriculture can be grouped into two categories: reducing and/or displacing emissions and enhancing carbon removal. Some of these reductions involve increasing the efficiency of farm operations (e.g. more fuel-efficient equipment) while some involve interruptions in the natural carbon cycle. Also, some effective techniques (such as the elimination of stubble burning) can negatively impact other environmental concerns (increased herbicide use to control weeds not destroyed by burning).

Reducing Emissions

Increasing yields and efficiency generally reduces emissions as well, since more food results from the same or less effort. Techniques include more accurate use of fertilizers, less soil disturbance, better irrigation, and crop strains bred for locally beneficial traits and increased yields.

Replacing more energy intensive farming operations can also reduce emissions. Reduced or no-till farming requires less machine use and burns correspondingly less fuel per acre. However, no-till usually increases use of weed-control chemicals and the residue now left on the soil surface is more likely to release its CO_2 to the atmosphere as it decays, reducing the net carbon reduction.

In practice, most farming operations that incorporate post-harvest crop residues, wastes and byproducts back into the soil provide a carbon storage benefit. This is particularly the case for practices such as field burning of stubble - rather than releasing almost all of the stored CO_2 to the atmosphere, tillage incorporates the biomass back into the soil.

Enhancing Carbon Removal

All crops absorb CO_2 during growth and release it after harvest. The goal of agricultural carbon removal is to use the crop and its relation to the carbon cycle to permanently sequester carbon within the soil. This is done by selecting farming methods that return biomass to the soil and enhance the conditions in which the carbon within the plants will be reduced to its elemental nature and stored in a stable state. Methods for accomplishing this include:

- Use cover crops such as grasses and weeds as temporary cover between planting seasons
- Concentrate livestock in small paddocks for days at a time so they graze lightly but evenly. This encourages roots to grow deeper into the soil. Stock also till the soil with their hooves, grinding old grass and manures into the soil.
- Cover bare paddocks with hay or dead vegetation. This protects soil from the sun and allows the soil to hold more water and be more attractive to carbon-capturing microbes.
- Restore degraded land, which slows carbon release while returning the land to agriculture or other use.

Agricultural sequestration practices may have positive effects on soil, air, and water quality, be beneficial to wildlife, and expand food production. On degraded croplands, an increase of 1 ton of soil carbon pool may increase crop yield by 20 to 40 kilograms per hectare of wheat, 10 to 20 kg/ ha for maize, and 0.5 to 1 kg/ha for cowpeas.

The effects of soil sequestration can be reversed. If the soil is disrupted or tillage practices are abandoned, the soil becomes a net source of greenhouse gases. Typically after 15 to 30 years of sequestration, soil becomes saturated and ceases to absorb carbon. This implies that there is a global limit to the amount of carbon that soil can hold.

Many factors affect the costs of carbon sequestration including soil quality, transaction costs and various externalities such as leakage and unforeseen environmental damage. Because reduction of atmospheric CO_2 is a long-term concern, farmers can be reluctant to adopt more expensive agricultural techniques when there is not a clear crop, soil, or economic benefit. Governments such as Australia and New Zealand are considering allowing farmers to sell carbon credits once they document that they have sufficiently increased soil carbon content.

Ocean-related

Iron Fertilization

Ocean iron fertilization is an example of such a geoengineering technique. Iron fertilization attempts to encourage phytoplankton growth, which removes carbon from the atmosphere for at least a period of time. This technique is controversial due to limited understanding of its complete effects on the marine ecosystem, including side effects and possibly large deviations from expected behavior. Such effects potentially include release of nitrogen oxides, and disruption of the ocean's nutrient balance.

Natural iron fertilisation events (e.g., deposition of iron-rich dust into ocean waters) can enhance carbon sequestration. Sperm whales act as agents of iron fertilisation when they transport iron from the deep ocean to the surface during prey consumption and defecation. Sperm whales have been shown to increase the levels of primary production and carbon export to the deep ocean by depositing iron rich feces into surface waters of the Southern Ocean. The iron rich feces causes phytoplankton to grow and take up more carbon from the atmosphere. When the phytoplankton dies, some of it sinks to the deep ocean and takes the atmospheric carbon with it. By reducing the abundance of sperm whales in the Southern Ocean, whaling has resulted in an extra 200,000 tonnes of carbon remaining in the atmosphere each year.

Urea Fertilization

Ian Jones proposes fertilizing the ocean with urea, a nitrogen rich substance, to encourage phytoplankton growth.

Australian company Ocean Nourishment Corporation (ONC) plans to sink hundreds of tonnes of urea into the ocean to boost CO_2-absorbing phytoplankton growth as a way to combat climate change. In 2007, Sydney-based ONC completed an experiment involving 1 tonne of nitrogen in the Sulu Sea off the Philippines.

Mixing Layers

Encouraging various ocean layers to mix can move nutrients and dissolved gases around, offering avenues for geoengineering. Mixing may be achieved by placing large vertical pipes in the oceans to pump nutrient rich water to the surface, triggering blooms of algae, which store carbon when they grow and export carbon when they die. This produces results somewhat similar to iron fertilization. One side-effect is a short-term rise in CO_2, which limits its attractiveness.

Seaweed

Seaweed grows very fast and can theoretically be harvested and processed to generate biomethane, via Anaerobic Digestion to generate electricity, via Cogeneration/CHP or as a replacement for natural gas. One study suggested that if seaweed farms covered 9% of the ocean they could produce enough biomethane to supply Earth's equivalent demand for fossil fuel energy, remove 53 gigatonnes of CO_2 per year from the atmosphere and sustainably produce 200 kg per year of fish, per person, for 10 billion people. Ideal species for such farming and conversion include Laminaria digitata, Fucus serratus and Saccharina latissima.

Physical Processes

Biochar can be landfilled, used as a soil improver or burned using carbon capture and storage

Biomass-related

Bio-energy with Carbon Capture and Storage (BECCS)

BECCS refers to biomass in power stations and boilers that use carbon capture and storage. The carbon sequestered by the biomass would be captured and stored, thus removing carbon dioxide from the atmosphere.

This technology is sometimes referred to as bio-energy with carbon storage, BECS, though this term can also refer to the carbon sequestration potential in other technologies, such as biochar.

Burial

Burying biomass (such as trees) directly, mimics the natural processes that created fossil fuels. Landfills also represent a physical method of sequestration.

Biochar Burial

Biochar is charcoal created by pyrolysis of biomass waste. The resulting material is added to a landfill or used as a soil improver to create terra preta. Addition of pyrogenic organic carbon (biochar) is a novel strategy to increase the soil-C stock for the long-term and to mitigate global-warming by offsetting the atmospheric C (up to 9.5 Pg C annually).

In the soil, the carbon is unavailable for oxidation to CO2 and consequential atmospheric release. This is one technique advocated by scientist James Lovelock, creator of the Gaia hypothesis. According to Simon Shackley, "people are talking more about something in the range of one to two billion tonnes a year."

The mechanisms related to biochar are referred to as bio-energy with carbon storage, BECS.

Ocean Storage

If CO_2 were to be injected to the ocean bottom, the pressures would be great enough for CO_2 to be in its liquid phase. The idea behind ocean injection would be to have stable, stationary pools of CO_2 at the ocean floor. The ocean could potentially hold over a thousand billion tons of CO_2. However, this avenue of sequestration isn't being as actively pursued because of concerns about the impact on ocean life, and concerns about its stability.

River mouths bring large quantities of nutrients and dead material from upriver into the ocean as part of the process that eventually produces fossil fuels. Transporting material such as crop waste out to sea and allowing it to sink exploits this idea to increase carbon storage. International regulations on marine dumping may restrict or prevent use of this technique.

Geological Sequestration

Geological sequestration refers to the storage of CO_2 underground in depleted oil and gas reservoirs, saline formations, or deep, un-minable coal beds.

Once CO_2 is captured from a gas or coal-fired power plant, it would be compressed to ≈100 bar so that it would be a supercritical fluid. In this fluid form, the CO_2 would be easy to transport via pipeline to the place of storage. The CO_2 would then be injected deep underground, typically around 1 km, where it would be stable for hundreds to millions of years. At these storage conditions, the density of supercritical CO_2 is 600 to 800 kg / m^3. For consumers, the cost of electricity from a coal-fired power plant with carbon capture and storage (CCS) is estimated to be 0.01 - 0.05 $ / kWh higher than without CCS. For reference, the average cost of electricity in the US in 2004 was 0.0762 $ / kWh. In other terms, the cost of CCS would be 20 - 70 $/ton of CO_2 captured. The transportation and injection of CO_2 is relatively cheap, with the capture costs accounting for 70 - 80% of CCS costs.

The important parameters in determining a good site for carbon storage are: rock porosity, rock permeability, absence of faults, and geometry of rock layers. The medium in which the CO_2 is to be stored ideally has a high porosity and permeability, such as sandstone or limestone. Sandstone can have a permeability ranging from 1 to 10^{-5} Darcy, and can have a porosity as high as ≈30%. The porous rock must be capped by a layer of low permeability which acts as a seal, or caprock, for the CO_2. Shale is an example of a very good caprock, with a permeability of 10^{-5} to 10^{-9} Darcy. Once injected, the CO_2 plume will rise via buoyant forces, since it is less dense than its surroundings. Once it encounters a caprock, it will spread laterally until it encounters a gap. If there are fault planes near the injection zone, there is a possibility the CO_2 could migrate along the fault to the surface, leaking into the atmosphere, which would be potentially dangerous to life in the surrounding area. Another danger related to carbon sequestration is induced seismicity. If the injection of CO_2 creates pressures that are too high underground, the formation will fracture, causing an earthquake.

While trapped in a rock formation, CO_2 can be in the supercritical fluid phase or dissolve in groundwater/brine. It can also react with minerals in the geologic formation to precipitate carbonates.

Worldwide storage capacity in oil and gas reservoirs is estimated to be 675 - 900 Gt CO_2, and in un-minable coal seams is estimated to be 15 - 200 Gt CO_2. Deep saline formations have the largest capacity, which is estimated to be 1,000 - 10,000 Gt CO_2. In the US, there is an estimated 160 Gt CO_2 storage capacity.

There are a number of large-scale carbon capture and sequestration projects that have demonstrated the viability and safety of this method of carbon storage, which are summarized here by the Global CCS Institute. The dominant monitoring technique is seismic imaging, where vibrations are generated that propagate through the subsurface. The geologic structure can be imaged from the refracted/reflected waves.

The first large-scale CO2 sequestration project which began in 1996 is called Sleipner, and is located in the North Sea where Norway's StatoilHydro strips carbon dioxide from natural gas with amine solvents and disposed of this carbon dioxide in a deep saline aquifer. In 2000, a coal-fueled synthetic natural gas plant in Beulah, North Dakota, became the world's first coal-using plant to capture and store carbon dioxide, at the Weyburn-Midale Carbon Dioxide Project.

CO2 has been used extensively in enhanced crude oil recovery operations in the United States beginning in 1972. There are in excess of 10,000 wells that inject CO2 in the state of Texas alone. The gas comes in part from anthropogenic sources, but is principally from large naturally occurring geologic formations of CO2. It is transported to the oil-producing fields through a large network of over 5,000 kilometres (3,100 mi) of CO2 pipelines. The use of CO2 for enhanced oil recovery (EOR) methods in heavy oil reservoirs in the Western Canadian Sedimentary Basin (WCSB) has also been proposed. However, transport cost remains an important hurdle. An extensive CO2 pipeline system does not yet exist in the WCSB. Athabasca oil sands mining that produces CO2 is hundreds of kilometers north of the subsurface Heavy crude oil reservoirs that could most benefit from CO2 injection.

Chemical Processes

Developed in the Netherlands, an electrocatalysis by a copper complex helps reduce carbon dioxide to oxalic acid; This conversion uses carbon dioxide as a feedstock to generate oxalic acid.

Mineral Carbonation

Carbon, in the form of CO_2 can be removed from the atmosphere by chemical processes, and stored in stable carbonate mineral forms. This process is known as 'carbon sequestration by mineral carbonation' or mineral sequestration. The process involves reacting carbon dioxide with abundantly available metal oxides–either magnesium oxide (MgO) or calcium oxide (CaO)–to form stable carbonates. These reactions are exothermic and occur naturally (e.g., the weathering of rock over geologic time periods).

$$CaO + CO_2 \rightarrow CaCO_3MgO + CO_2 \rightarrow MgCO_3$$

Calcium and magnesium are found in nature typically as calcium and magnesium silicates (such as forsterite and serpentinite) and not as binary oxides. For forsterite and serpentine the reactions are:

$$Mg_2SiO_4 + 2\ CO_2 \rightarrow 2\ MgCO_3 + SiO_2$$

$$Mg_3Si_2O_5(OH)_4 + 3\ CO_2 \rightarrow 3\ MgCO_3 + 2\ SiO_2 + 2\ H_2O$$

The following table lists principal metal oxides of Earth's crust. Theoretically up to 22% of this mineral mass is able to form carbonates.

Earthen Oxide	Percent of Crust	Carbonate	Enthalpy change (kJ/mol)
SiO_2	59.71		
Al_2O_3	15.41		
CaO	4.90	$CaCO_3$	-179
MgO	4.36	$MgCO_3$	-117
Na_2O	3.55	Na_2CO_3	
FeO	3.52	$FeCO_3$	
K_2O	2.80	K_2CO_3	
Fe_2O_3	2.63	$FeCO_3$	
	21.76	All Carbonates	

These reactions are slightly more favorable at low temperatures. This process occurs naturally over geologic time frames and is responsible for much of the Earth's surface limestone. The reaction rate can be made faster, for example by reacting at higher temperatures and/or pressures, or by pre-treatment, although this method requires additional energy. Experiments suggest this process is reasonably quick (one year) given porous basaltic rocks.

CO2 naturally reacts with peridotite rock in surface exposures of ophiolites, notably in Oman. It has been suggested that this process can be enhanced to carry out natural mineralisation of CO2.

Industrial Use

Traditional cement manufacture releases large amounts of carbon dioxide, but newly developed cement types from Novacem can absorb CO2 from ambient air during hardening. A similar technique was pioneered by TecEco, which has been producing "EcoCement" since 2002.

In Estonia, oil shale ash, generated by power stations could be used as sorbents for CO2 mineral sequestration. The amount of CO2 captured averaged 60 to 65% of the carbonaceous CO2 and 10 to 11% of the total CO2 emissions.

Chemical Scrubbers

Various carbon dioxide scrubbing processes have been proposed to remove CO from the air, usually using a variant of the Kraft process. Carbon dioxide scrubbing variants exist based on potassium carbonate, which can be used to create liquid fuels, or on sodium hydroxide. These notably include artificial trees proposed by Klaus Lackner to remove carbon dioxide from the atmosphere using chemical scrubbers.

Ocean-related

Basalt Storage

Carbon dioxide sequestration in basalt involves the injecting of CO2 into deep-sea formations. The CO2 first mixes with seawater and then reacts with the basalt, both of which are alkaline-rich elements. This reaction results in the release of Ca^{2+} and Mg^{2+} ions forming stable carbonate minerals.

Underwater basalt offers a good alternative to other forms of oceanic carbon storage because it has a number of trapping measures to ensure added protection against leakage. These measures include "geothermal, sediment, gravitational and hydrate formation." Because CO2 hydrate is denser than CO2 in seawater, the risk of leakage is minimal. Injecting the CO2 at depths greater than 2,700 meters (8,900 ft) ensures that the CO2 has a greater density than seawater, causing it to sink.

One possible injection site is Juan de Fuca plate. Researchers at the Lamont-Doherty Earth Observatory found that this plate at the western coast of the United States has a possible storage capacity of 208 gigatons. This could cover the entire current U.S.

carbon emissions for over 100 years.

This process is undergoing tests as part of the CarbFix project, resulting in 95% of the injected 250 tonnes of CO2 to solidify into calcite in 2 years, using 25 tonnes of water per tonne of CO2.

Acid Neutralisation

Carbon dioxide forms carbonic acid when dissolved in water, so ocean acidification is a significant consequence of elevated carbon dioxide levels, and limits the rate at which it can be absorbed into the ocean (the solubility pump). A variety of different bases have been suggested that could neutralize the acid and thus increase CO2 absorption. For example, adding crushed limestone to oceans enhances the absorption of carbon dioxide. Another approach is to add sodium hydroxide to oceans which is produced by electrolysis of salt water or brine, while eliminating the waste hydrochloric acid by reaction with a volcanic silicate rock such as enstatite, effectively increasing the rate of natural weathering of these rocks to restore ocean pH.

Obstruction

Danger of Leaks

Carbon dioxide may be stored deep underground. At depth, hydrostatic pressure acts to keep it in a liquid state. Reservoir design faults, rock fissures and tectonic processes may act to release the gas stored into the ocean or atmosphere.

Financial Costs

The use of the technology would add an additional 1-5 cents of cost per kilowatt hour, according to estimate made by the Intergovernmental Panel on Climate Change. The financial costs of modern coal technology would nearly double if use of CCS technology were to be required by regulation. The cost of CCS technology differs with the different types of capture technologies being used and with the different sites that it is implemented in, but the costs tend to increase with CCS capture implementation. One study conducted predicted that with new technologies these costs could be lowered but would remain slightly higher than prices without CCS technologies.

Energy Requirements

The energy requirements of sequestration processes may be significant. In one paper, sequestration consumed 25 percent of the plant's rated 600 megawatt output capacity.

> *After adding CO_2 capture and compression, the capacity of the coal-fired power plant is reduced to 457 MW.*

Eutrophication

The eutrophication of the Potomac River is evident from the bright green water, caused by a dense bloom of cyanobacteria.

Eutrophication is the depletion of oxygen in a water body, which kills aquatic animals. It is a response to the addition of excess nutrients, mainly phosphates, which induces explosive growth of plants and algae, the decaying of which consumes oxygen from the water. One example is the “bloom” or great increase of phytoplankton in a water body as a response to increased levels of nutrients. Eutrophication is almost always induced by the discharge of phosphate-containing detergents, fertilizers, or sewage, into an aquatic system.

Mechanism of Eutrophication

Eutrophication arises from the oversupply of nutrients, which leads to over growth of plants and algae. After such organisms die, the bacterial degradation of their biomass consumes the oxygen in the water, thereby creating the state of hypoxia.

According to Ullmann’s Encyclopedia, “the primary limiting factor for eutrophication is phosphate.” The availability of phosphorus generally promotes excessive plant growth and decay, favouring simple algae and plankton over other more complicated plants, and causes a severe reduction in water quality. Phosphorus is a necessary nutrient for plants to live, and is the limiting factor for plant growth in many freshwater ecosystems. Phosphate adheres tightly to soil, so it is mainly transported by erosion. Once translocated to lakes, the extraction of phosphate into water is slow, hence the difficulty of reversing the effects of eutrophication.

The sources of these excess phosphates are detergents, industrial/domestic run-offs, and fertilizers. With the phasing out of phosphate-containing detergents in the 1970s, industrial/domestic run-off and agriculture have emerged as the dominant contributors to eutrophication.

$$\left[{}^{-}O-\overset{\overset{O}{\|}}{\underset{\underset{O^{-}}{|}}{P}}-O-\overset{\overset{O}{\|}}{\underset{\underset{O^{-}}{|}}{P}}-O-\overset{\overset{O}{\|}}{\underset{\underset{O^{-}}{|}}{P}}-O^{-} \right] \left[Na^{+} \right]_5$$

Sodium triphosphate, once a component of many detergents, was a major contributor to eutrophication.

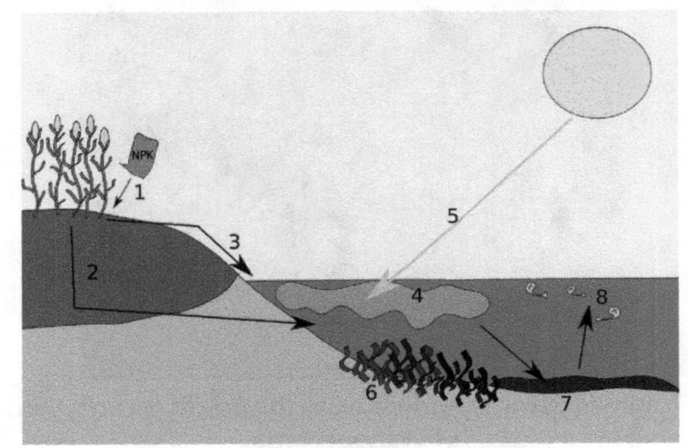

1. Excess nutrients are applied to the soil. 2. Some nutrients leach into the soil where they can remain for years. Eventually, they get drained into the water body. 3. Some nutrients run off over the ground into the body of water. 4. The excess nutrients cause an algal bloom. 5. The algal bloom blocks the light of the sun from reaching the bottom of the water body. 6. The plants beneath the algal bloom die because they cannot get sunlight to photosynthesize. 7. Eventually, the algal bloom dies and sinks to the bottom of the lake. Bacteria begins to decompose the remains, using up oxygen for respiration. 8. The decomposition causes the water to become depleted of oxygen. Larger life forms, such as fish, suffocate to death. This body of water can no longer support life.

Lakes and Rivers

Eutrophication in a canal

When algae die, they decompose and the nutrients contained in that organic matter are converted into inorganic form by microorganisms. This decomposition process consumes oxygen, which reduces the concentration of dissolved oxygen. The depleted oxygen levels in turn may lead to fish kills and a range of other effects reducing biodiversity. Nutrients may become concentrated in an anoxic zone and may only be made available again during autumn turn-over or in conditions of turbulent flow.

Enhanced growth of aquatic vegetation or phytoplankton and algal blooms disrupts normal functioning of the ecosystem, causing a variety of problems such as a lack of oxygen needed for fish and shellfish to survive. The water becomes cloudy, typically coloured a shade of green, yellow, brown, or red. Eutrophication also decreases the value of rivers, lakes and aesthetic enjoyment. Health problems can occur where eutrophic conditions interfere with drinking water treatment.

Human activities can accelerate the rate at which nutrients enter ecosystems. Runoff from agriculture and development, pollution from septic systems and sewers, sewage sludge spreading, and other human-related activities increase the flow of both inorganic nutrients and organic substances into ecosystems. Elevated levels of atmospheric compounds of nitrogen can increase nitrogen availability. Phosphorus is often regarded as the main culprit in cases of eutrophication in lakes subjected to "point source" pollution from sewage pipes. The concentration of algae and the trophic state of lakes correspond well to phosphorus levels in water. Studies conducted in the Experimental Lakes Area in Ontario have shown a relationship between the addition of phosphorus and the rate of eutrophication. Humankind has increased the rate of phosphorus cycling on Earth by four times, mainly due to agricultural fertilizer production and application. Between 1950 and 1995, an estimated 600,000,000 tonnes of phosphorus were applied to Earth's surface, primarily on croplands. Policy changes to control point sources of phosphorus have resulted in rapid control of eutrophication.

Natural Eutrophication

Although eutrophication is commonly caused by human activities, it can also be a natural process, particularly in lakes. Eutrophy occurs in many lakes in temperate grasslands, for instance. Paleolimnologists now recognise that climate change, geology, and other external influences are critical in regulating the natural productivity of lakes. Some lakes also demonstrate the reverse process (meiotrophication), becoming less nutrient rich with time. The main difference between natural and anthropogenic eutrophication is that the natural process is very slow, occurring on geological time scales.

Ocean Waters

Eutrophication is a common phenomenon in coastal waters. In contrast to freshwater systems, nitrogen is more commonly the key limiting nutrient of marine waters; thus, nitrogen levels have greater importance to understanding eutrophication problems in salt water. Estuaries tend to be naturally eutrophic because land-derived nutrients are concentrated where run-off enters a confined channel. Upwelling in coastal systems also promotes increased productivity by conveying deep, nutrient-rich waters to the surface, where the nutrients can be assimilated by algae.

The World Resources Institute has identified 375 hypoxic coastal zones in the world, concentrated in coastal areas in Western Europe, the Eastern and Southern coasts of the US, and East Asia, particularly Japan.

In addition to runoff from land, atmospheric fixed nitrogen can enter the open ocean. A study in 2008 found that this could account for around one third of the ocean's external (non-recycled) nitrogen supply, and up to 3% of the annual new marine biological production. It has been suggested that accumulating reactive nitrogen in the environment may prove as serious as putting carbon dioxide in the atmosphere.

Terrestrial Ecosystems

Terrestrial ecosystems are subject to similarly adverse impacts from eutrophication. Increased nitrates in soil are frequently undesirable for plants. Many terrestrial plant species are endangered as a result of soil eutrophication, such as the majority of orchid species in Europe. Meadows, forests, and bogs are characterized by low nutrient content and slowly growing species adapted to those levels, so they can be overgrown by faster growing and more competitive species. In meadows, tall grasses that can take advantage of higher nitrogen levels may change the area so that natural species may be lost. Species-rich fens can be overtaken by reed or reedgrass species. Forest undergrowth affected by run-off from a nearby fertilized field can be turned into a nettle and bramble thicket.

Chemical forms of nitrogen are most often of concern with regard to eutrophication, because plants have high nitrogen requirements so that additions of nitrogen compounds will stimulate plant growth. Nitrogen is not readily available in soil because N_2, a gaseous form of nitrogen, is very stable and unavailable directly to higher plants. Terrestrial ecosystems rely on microbial nitrogen fixation to convert N_2 into other forms such as nitrates. However, there is a limit to how much nitrogen can be utilized. Ecosystems receiving more nitrogen than the plants require are called nitrogen-saturated. Saturated terrestrial ecosystems then can contribute both inorganic and organic nitrogen to freshwater, coastal, and marine eutrophication, where nitrogen is also typically a limiting nutrient. This is also the case with increased levels of phosphorus. However, because phosphorus is generally much less soluble than nitrogen, it is leached from the soil at a much slower rate than nitrogen. Consequently, phosphorus is much more important as a limiting nutrient in aquatic systems.

Ecological Effects

Eutrophication was recognized as a water pollution problem in European and North American lakes and reservoirs in the mid-20th century. Since then, it has become more widespread. Surveys showed that 54% of lakes in Asia are eutrophic; in Europe, 53%; in North America, 48%; in South America, 41%; and in Africa, 28%.

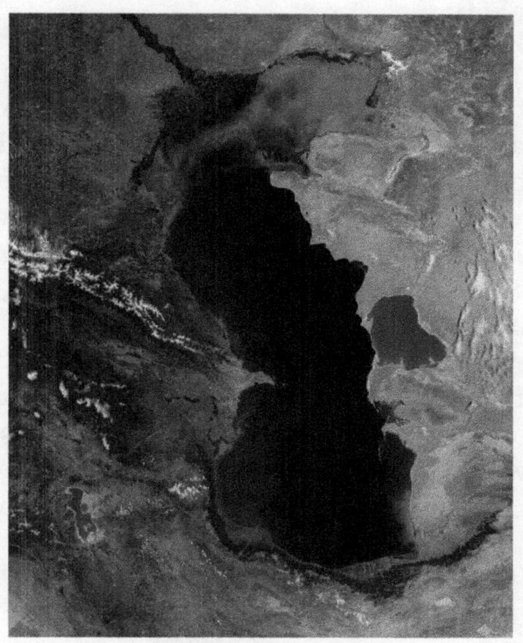

Eutrophication is apparent as increased turbidity in the northern part of the Caspian Sea, imaged from orbit.

Many ecological effects can arise from stimulating primary production, but there are three particularly troubling ecological impacts: decreased biodiversity, changes in species composition and dominance, and toxicity effects.

- Increased biomass of phytoplankton
- Toxic or inedible phytoplankton species
- Increases in blooms of gelatinous zooplankton
- Increased biomass of benthic and epiphytic algae
- Changes in macrophyte species composition and biomass
- Decreases in water transparency (increased turbidity)
- Colour, smell, and water treatment problems
- Dissolved oxygen depletion
- Increased incidences of fish kills
- Loss of desirable fish species
- Reductions in harvestable fish and shellfish

- Decreases in perceived aesthetic value of the water body

Decreased Biodiversity

When an ecosystem experiences an increase in nutrients, primary producers reap the benefits first. In aquatic ecosystems, species such as algae experience a population increase (called an algal bloom). Algal blooms limit the sunlight available to bottom-dwelling organisms and cause wide swings in the amount of dissolved oxygen in the water. Oxygen is required by all aerobically respiring plants and animals and it is replenished in daylight by photosynthesizing plants and algae. Under eutrophic conditions, dissolved oxygen greatly increases during the day, but is greatly reduced after dark by the respiring algae and by microorganisms that feed on the increasing mass of dead algae. When dissolved oxygen levels decline to hypoxic levels, fish and other marine animals suffocate. As a result, creatures such as fish, shrimp, and especially immobile bottom dwellers die off. In extreme cases, anaerobic conditions ensue, promoting growth of bacteria such as *Clostridium botulinum* that produces toxins deadly to birds and mammals. Zones where this occurs are known as dead zones.

New Species Invasion

Eutrophication may cause competitive release by making abundant a normally limiting nutrient. This process causes shifts in the species composition of ecosystems. For instance, an increase in nitrogen might allow new, competitive species to invade and out-compete original inhabitant species. This has been shown to occur in New England salt marshes. In Europe and Asia, the Common carp frequently lives in naturally Eutrophic or Hypereutrophic areas, and is adapted to living in such conditions. The eutrophication of areas outside its natural range partially explain the fish's success in colonising these areas after being introduced.

Toxicity

Some algal blooms, otherwise called "nuisance algae" or "harmful algal blooms", are toxic to plants and animals. Toxic compounds they produce can make their way up the food chain, resulting in animal mortality. Freshwater algal blooms can pose a threat to livestock. When the algae die or are eaten, neuro- and hepatotoxins are released which can kill animals and may pose a threat to humans. An example of algal toxins working their way into humans is the case of shellfish poisoning. Biotoxins created during algal blooms are taken up by shellfish (mussels, oysters), leading to these human foods acquiring the toxicity and poisoning humans. Examples include paralytic, neurotoxic, and diarrhoetic shellfish poisoning. Other marine animals can be vectors for such toxins, as in the case of ciguatera, where it is typically a predator fish that accumulates the toxin and then poisons humans.

Sources of High Nutrient Runoff

Characteristics of point and nonpoint sources of chemical inputs modified from Novonty and Olem 1994)
Point sources • Wastewater effluent (municipal and industrial) • Runoff and leachate from waste disposal systems • Runoff and infiltration from animal feedlots • Runoff from mines, oil fields, unsewered industrial sites • Overflows of combined storm and sanitary sewers • Runoff from construction sites less than 20,000 m^2 (220,000 ft^2) • Untreated sewage
Nonpoint sources • Runoff from agriculture/irrigation • Runoff from pasture and range • Urban runoff from unsewered areas • Septic tank leachate • Runoff from construction sites >20,000 m^2 • Runoff from abandoned mines • Atmospheric deposition over a water surface • Other land activities generating contaminants

In order to gauge how to best prevent eutrophication from occurring, specific sources that contribute to nutrient loading must be identified. There are two common sources of nutrients and organic matter: point and nonpoint sources.

Point Sources

Point sources are directly attributable to one influence. In point sources the nutrient waste travels directly from source to water. Point sources are relatively easy to regulate.

Nonpoint Sources

Nonpoint source pollution (also known as 'diffuse' or 'runoff' pollution) is that which comes from ill-defined and diffuse sources. Nonpoint sources are difficult to regulate and usually vary spatially and temporally (with season, precipitation, and other irregular events).

It has been shown that nitrogen transport is correlated with various indices of human activity in watersheds, including the amount of development. Ploughing in agriculture and development are activities that contribute most to nutrient loading. There are three reasons that nonpoint sources are especially troublesome:

Soil Retention

Nutrients from human activities tend to accumulate in soils and remain there for years. It has been shown that the amount of phosphorus lost to surface waters increases linearly with the amount of phosphorus in the soil. Thus much of the nutrient loading in soil eventually makes its way to water. Nitrogen, similarly, has a turnover time of decades.

Runoff to Surface Water and Leaching to Groundwater

Nutrients from human activities tend to travel from land to either surface or ground water. Nitrogen in particular is removed through storm drains, sewage pipes, and other forms of surface runoff. Nutrient losses in runoff and leachate are often associated with agriculture. Modern agriculture often involves the application of nutrients onto fields in order to maximise production. However, farmers frequently apply more nutrients than are taken up by crops or pastures. Regulations aimed at minimising nutrient exports from agriculture are typically far less stringent than those placed on sewage treatment plants and other point source polluters. It should be also noted that lakes within forested land are also under surface runoff influences. Runoff can wash out the mineral nitrogen and phosphorus from detritus and in consequence supply the water bodies leading to slow, natural eutrophication.

Atmospheric Deposition

Nitrogen is released into the air because of ammonia volatilization and nitrous oxide production. The combustion of fossil fuels is a large human-initiated contributor to atmospheric nitrogen pollution. Atmospheric deposition (e.g., in the form of acid rain) can also affect nutrient concentration in water, especially in highly industrialized regions.

Other Causes

Any factor that causes increased nutrient concentrations can potentially lead to eutrophication. In modeling eutrophication, the rate of water renewal plays a critical role; stagnant water is allowed to collect more nutrients than bodies with replenished water supplies. It has also been shown that the drying of wetlands causes an increase in nutrient concentration and subsequent eutrophication blooms.

Prevention and Reversal

Eutrophication poses a problem not only to ecosystems, but to humans as well. Reducing eutrophication should be a key concern when considering future policy, and a sustainable solution for everyone, including farmers and ranchers, seems feasible. While eutrophication does pose problems, humans should be aware that natural runoff (which causes algal blooms in the wild) is common in ecosystems and should thus not reverse nutrient concentrations beyond normal levels. Cleanup measures have been mostly, but not completely, successful. Finnish phosphorus removal measures started in the mid-1970s and have targeted rivers and lakes polluted by industrial and municipal discharges. These efforts have had a 90% removal efficiency. Still, some targeted point sources did not show a decrease in runoff despite reduction efforts.

Shellfish in Estuaries: Unique Solutions

One proposed solution to eutrophication in estuaries is to restore shellfish populations, such as oysters and mussels. Oyster reefs remove nitrogen from the water column and filter out suspended solids, subsequently reducing the likelihood or extent of harmful algal blooms or anoxic conditions. Filter feeding activity is considered beneficial to water quality by controlling phytoplankton density and sequestering nutrients, which can be removed from the system through shellfish harvest, buried in the sediments, or lost through denitrification. Foundational work toward the idea of improving marine water quality through shellfish cultivation was conducted by Odd Lindahl et al., using mussels in Sweden. In the United States, shellfish restoration projects have been conducted on the East, West and Gulf coasts.

Minimizing Nonpoint Pollution: Future Work

Nonpoint pollution is the most difficult source of nutrients to manage. The literature suggests, though, that when these sources are controlled, eutrophication decreases. The following steps are recommended to minimize the amount of pollution that can enter aquatic ecosystems from ambiguous sources.

Riparian Buffer Zones

Studies show that intercepting non-point pollution between the source and the water is a successful means of prevention. Riparian buffer zones are interfaces between a flowing body of water and land, and have been created near waterways in an attempt to filter pollutants; sediment and nutrients are deposited here instead of in water. Creating buffer zones near farms and roads is another possible way to prevent nutrients from traveling too far. Still, studies have shown that the effects of atmospheric nitrogen pollution can reach far past the buffer zone. This suggests that the most effective means of prevention is from the primary source.

Prevention Policy

Laws regulating the discharge and treatment of sewage have led to dramatic nutrient reductions to surrounding ecosystems, but it is generally agreed that a policy regulating agricultural use of fertilizer and animal waste must be imposed. In Japan the amount of nitrogen produced by livestock is adequate to serve the fertilizer needs for the agriculture industry. Thus, it is not unreasonable to command livestock owners to clean up animal waste—which when left stagnant will leach into ground water.

Policy concerning the prevention and reduction of eutrophication can be broken down into four sectors: Technologies, public participation, economic instruments, and cooperation. The term technology is used loosely, referring to a more widespread use of existing methods rather than an appropriation of new technologies. As mentioned before, nonpoint sources of pollution are the primary contributors to eutrophication, and their effects can be easily minimized through common agricultural practices. Reducing the amount of pollutants that reach a watershed can be achieved through the protection of its forest cover, reducing the amount of erosion leeching into a watershed. Also, through the efficient, controlled use of land using sustainable agricultural practices to minimize land degradation, the amount of soil runoff and nitrogen-based fertilizers reaching a watershed can be reduced. Waste disposal technology constitutes another factor in eutrophication prevention. Because a major contributor to the nonpoint source nutrient loading of water bodies is untreated domestic sewage, it is necessary to provide treatment facilities to highly urbanized areas, particularly those in underdeveloped nations, in which treatment of domestic waste water is a scarcity. The technology to safely and efficiently reuse waste water, both from domestic and industrial sources, should be a primary concern for policy regarding eutrophication.

The role of the public is a major factor for the effective prevention of eutrophication. In order for a policy to have any effect, the public must be aware of their contribution to the problem, and ways in which they can reduce their effects. Programs instituted to promote participation in the recycling and elimination of wastes, as well as education on the issue of rational water use are necessary to protect water quality within urbanized areas and adjacent water bodies.

Economic instruments, "which include, among others, property rights, water markets, fiscal and financial instruments, charge systems and liability systems, are gradually becoming a substantive component of the management tool set used for pollution control and water allocation decisions." Incentives for those who practice clean, renewable, water management technologies are an effective means of encouraging pollution prevention. By internalizing the costs associated with the negative effects on the environment, governments are able to encourage a cleaner water management.

Because a body of water can have an effect on a range of people reaching far beyond that of the watershed, cooperation between different organizations is necessary to prevent the intrusion of contaminants that can lead to eutrophication. Agencies ranging from state governments to those of water resource management and non-governmental organizations, going as low as the local population, are responsible for preventing eutrophication of water bodies. In the United States, the most well known inter-state effort to prevent eutrophication is the Chesapeake Bay.

Nitrogen Testing and Modeling

Soil Nitrogen Testing (N-Testing) is a technique that helps farmers optimize the amount of fertilizer applied to crops. By testing fields with this method, farmers saw a decrease in fertilizer application costs, a decrease in nitrogen lost to surrounding sources, or both. By testing the soil and modeling the bare minimum amount of fertilizer needed, farmers reap economic benefits while reducing pollution.

Organic Farming

There has been a study that found that organically fertilized fields "significantly reduce harmful nitrate leaching" over conventionally fertilized fields. However, a more recent study found that eutrophication impacts are in some cases higher from organic production than they are from conventional production.

Cultural Eutrophication

Cultural eutrophication is the process that speeds up natural eutrophication because of human activity. Due to clearing of land and building of towns and cities, land runoff is accelerated and more nutrients such as phosphates and nitrate are supplied to lakes and rivers, and then to coastal estuaries and bays. Extra nutrients are also supplied by treatment plants, golf courses, fertilizers, and farms.

These nutrients result in an excessive growth of plant life known as an algal bloom. This can change a lake's natural food web, and also reduce the amount of dissolved oxygen in the water for organisms to breathe. Both these things cause animal and plant death rates to increase as the plants take in poisonous water while the animals drink the poisoned water. This contaminates water, making it undrinkable, and sediment quickly fills the lake. Cultural eutrophication is a form of water pollution.

Cultural eutrophication also occurs when excessive fertilizers run into lakes and rivers. This encourages the growth of algae (algal bloom) and other aquatic plants. Following this, overcrowding occurs and plants compete for sunlight, space and oxygen. Overgrowth of water plants also blocks sunlight and oxygen for aquatic life in the water, which in turn threatens their survival. Algae also grows easily, thus threatening other water plants no matter whether they are floating, half-submerged, or fully submerged. Not only does this cause algal blooming, it can cause an array of more long-term effects on the water such as damage to coral reefs and deep sea animal life. It also speeds up the damage of both marine and also affects humans if the effects of algal blooming is too drastic. Fish will die and there will be lack of food in the area. Nutrient pollution is a major cause of algal blooming, and should be minimized.

The Experimental Lakes Area (ELA), Ontario, Canada is a fully equipped, year-round, permanent field station that uses the whole ecosystem approach and long-term, whole-lake investigations of freshwater focusing on cultural eutrophication. ELA is currently cosponsored by the Canadian Departments of Environment and Fisheries and Oceans, with a mandate to investigate the aquatic effects of a wide variety of stresses on lakes and their catchments

Continental Shelf Pump

In oceanic biogeochemistry, the continental shelf pump is proposed to operate in the shallow waters of the continental shelves, acting as a mechanism to transport carbon (as either dissolved or particulate material) from surface waters to the interior of the adjacent deep ocean.

Overview

Originally formulated by Tsunogai *et al.* (1999), the pump is believed to occur where the solubility and biological pumps interact with a local hydrography that feeds dense water from the shelf floor into sub-surface (at least subthermocline) waters in the neighbouring deep ocean. Tsunogai *et al.*'s (1999) original work focused on the East China Sea, and the observation that, averaged over the year, its surface waters represented a sink for carbon dioxide. This observation was combined with others of the distribution of dissolved carbonate and alkalinity and explained as follows :

- the shallowness of the continental shelf restricts convection of cooling water
- as a consequence, cooling is greater for continental shelf waters than for neighbouring open ocean waters
- this leads to the production of relatively cool and dense water on the shelf
- the cooler waters promote the solubility pump and lead to an increased storage of dissolved inorganic carbon
- this extra carbon storage is augmented by the increased biological production characteristic of shelves

- the dense, carbon-rich shelf waters sink to the shelf floor and enter the sub-surface layer of the open ocean via isopycnal mixing

Significance

Based on their measurements of the CO_2 flux over the East China Sea (35 g C m^{-2} y^{-1}), Tsunogai *et al.* (1999) estimated that the continental shelf pump could be responsible for an air-to-sea flux of approximately 1 Gt C y^{-1} over the world's shelf areas. Given that observational and modelling of anthropogenic emissions of CO_2 estimates suggest that the ocean is currently responsible for the uptake of approximately 2 Gt C y^{-1}, and that these estimates are poor for the shelf regions, the continental shelf pump may play an important role in the ocean's carbon cycle.

One caveat to this calculation is that the original work was concerned with the hydrography of the East China Sea, where cooling plays the dominant role in the formation of dense shelf water, and that this mechanism may not apply in other regions. However, it has been suggested that other processes may drive the pump under different climatic conditions. For instance, in polar regions, the formation of sea-ice results in the extrusion of salt that may increase seawater density. Similarly, in tropical regions, evaporation may increase local salinity and seawater density.

The strong sink of CO_2 at temperate latitudes reported by Tsunogai *et al.* (1999) was later confirmed in the Gulf of Biscay, the Middle Atlantic Bight and the North Sea. On the other hand, in the sub-tropical South Atlantic Bight reported a source of CO_2 to the atmosphere.

Recently, work has compiled and scaled available data on CO_2 fluxes in coastal environments, and shown that globally marginal seas act as a significant CO_2 sink (-1.6 mol C m^{-2} y^{-1}; -0.45 Gt C y^{-1}) in agreement with previous estimates. However, the global sink of CO_2 in marginal seas could be almost fully compensated by the emission of CO_2 (+11.1 mol C m^{-2} y^{-1}; +0.40 Gt C y^{-1}) from the ensemble of near-shore coastal ecosystems, mostly related to the emission of CO_2 from estuaries (0.34 Gt C y^{-1}).

An interesting application of this work has been examining the impact of sea level rise over the last de-glacial transition on the global carbon cycle. During the last glacial maximum sea level was some 120 m lower than today. As sea level rose the surface area of the shelf seas grew and in consequence the strength of the shelf sea pump should increase.

Remineralisation

In biogeochemistry, remineralisation (UK, US Spelling: remineralization) refers to the breakdown or transformation of organic matter (those molecules derived from a biological source) into its simplest inorganic forms. These transformations form a crucial link within ecosystems as they are responsible for liberating the energy stored in organic molecules and recycling matter within the system to be reused as nutrients by other organisms.

Remineralization is normally viewed as it relates to the cycling of the major biologically-important elements such as carbon, nitrogen and phosphorus. While crucial to all ecosystems, the process receives special consideration in aquatic settings, where it forms a significant link in the biogeochemical dynamics and cycling of aquatic ecosystems.

Role in Biogeochemistry

The term "remineralization" is used in several contexts across different disciplines. The term is most commonly used in the medicinal and physiological fields, where it is describes the development or redevelopment of mineralized structures in organisms such as teeth or bone. In the field of biogeochemistry, however, remineralization is used to describe a link in the chain of elemental cycling within a specific ecosystem. In particular, remineralization represents the point where organic material constructed by living organisms is broken down into basal inorganic components that are not obviously identifiable as having come from an organic source. This differs from the process of decomposition which is a more general descriptor of larger structures degrading to smaller structures.

Biogeochemists study this process across all ecosystems for a variety of reasons. This is done primarily to investigate the flow of material and energy in a given system, which is key to understanding the productivity of that ecosystem along with how it recycles material versus how much is entering the system. Understanding the rates and dynamics of organic matter remineralization in a given system can help in determining how or why some ecosystems might be more productive than others.

Remineralization Reactions

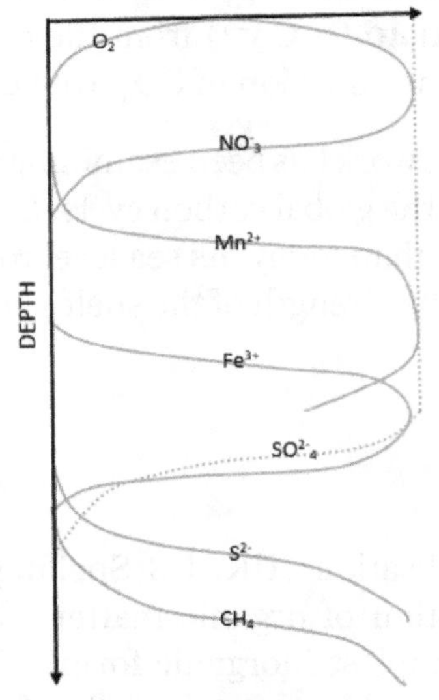

Sketch of major electron acceptors in marine sediment porewater based on idealized relative depths

While it is important to note that the process of remineralization is a series of complex biochemical pathways [within microbes], it can often be simplified as a series of one-step processes for ecosystem-level models and calculations. A generic form of these reactions is shown by:

$$Organic\ Matter + Oxidant -> Liberated\ Simple\ Nutrients + \underbrace{CO2}_{Carbon\ Dioxide} + \underbrace{H2O}_{Water}$$

The above generic equation starts with two reactants: some piece of organic matter (composed of organic carbon) and an oxidant. Most organic carbon exists in a reduced form which is then oxidized by the oxidant (such as O_2) into carbon dioxide and energy that can be harnessed by the organism. This process generally produces CO_2, water and a collection of simple nutrients like nitrate or phosphate that can then be taken up by other organisms. The above general form, when considering O_2 as the oxidant, is the equation for respiration. In this context specifically, the above equation represents bacterial respiration though the reactants and products are essentially analogous to the short-hand equations used for multi-cellular respiration.

Electron Acceptor Cascade

The degradation of organic matter through respiration in the modern ocean is facilitated by different electron acceptors, their favorability based on Gibbs free energy law, and the laws of thermodynamics. This redox chemistry is the basis for life in deep sea sediments and determines the obtainability of energy to organisms that live there. From the water interface moving toward deeper sediments, the order of these acceptors is oxygen, nitrate, manganese, iron, and sulfate. Moving downwards from the surface through the zonation of these deep ocean sediments, acceptors are used and depleted. Once depleted the next acceptor of lower favorability takes its place. Thermodynamically, oxygen represents the most favorable electron accepted but is quickly used up in the water sediment interface and O_2 concentrations extends only millimeters to centimeters down into the sediment in most locations of the deep sea. This favorability indicates an organism's ability to obtain higher energy from the reaction which helps them complete with other organisms. In the absence of these acceptors, organic matter can also be degraded through methanogenesis, but the net oxidation of this organic matter is not fully represented by this process. Each pathway and the stoichiometry of its reaction are listed in table 1.

Due to this quick depletion of O_2 in the surface sediments, a majority of microbes use anaerobic pathways to metabolize other oxides such as manganese, iron, and sulfate. It is also important to figure in bioturbation and the constant mixing of this material which can change the relative importance of each respiration pathway. For the microbial perspective please reference the electron transport chain.

Remineralization in Sediments

Reactions

A quarter of all organic material that exits the photic zone makes it to the seafloor without being remineralized and 90% of that remaining material is remineralized in sediments itself. Once in the sediment, organic remineralization may occur through a variety of reactions. The following reactions are the primary ways in which organic matter is remineralized, in them general organic matter (OM) is often represented by the shorthand: $(CH_2O)_{106}(NH_3)_{16}(H_3PO_4)$

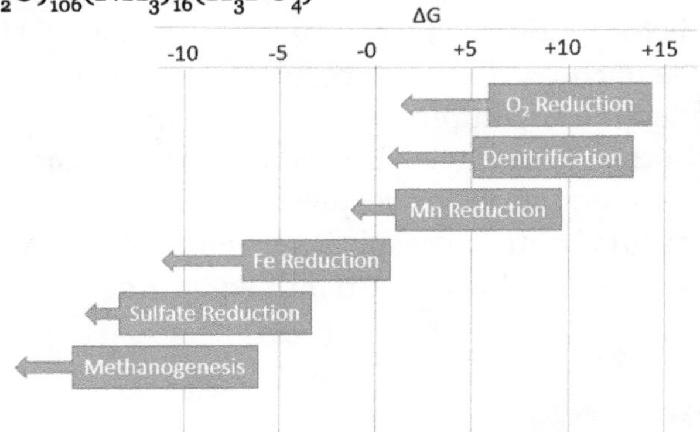

Relative favorability of reduction reactions in marine sediments based on thermodynamic energetics. Origin of arrows indicate energy associated with half-cell reaction. Length of arrow indicates an estimate of ΔG for the reaction (Adapted from Libes, 2011).

Aerobic Respiration

Aerobic respiration is the most preferred remineralization reaction due to its high energy yield. Although oxygen is quickly depleted in the sediments and is generally exhausted centimeters from the sediment-water interface.

Anaerobic Respiration

Respiration type	Reaction		ΔG
Aerobic	Oxygen reduction	$M + 150O2 -> 106CO2 + 16HNO3 + H3PO4 + 78H2O$	-29.9
	Denitrification	$OM + 104HNO3 -> 106CO2 + 60N2 + H3PO4 + 138H2O$	-28.4
	Manganese reduction	$OM + 260MnO2 + 174H2O -> 106CO2 + 8N2 + H3PO4 + 260Mn(OH)2$	-7.2
Anerobic	Iron reduction	$OM + 236Fe2O3 + 410H2O -> 106CO2 + 16NH3 + H3PO4 + 472Fe(OH)2$	-21.0
	Sulfate reduction	$OM + 59H2SO4 -> 106CO2 + 16NH3 + H3PO4 + 59H2S + 62H2O$	-6.1
	Methane fermentation (Methanogenesis)	$OM + 59H2O -> 47CO2 + 59CH4 + 16NH3 + H3PO4$	-5.6

In instances in which the environment is suboxic or anoxic, organisms will prefer to utilize denitrification to remineralize organic matter as it provides the second largest amount of energy. In depths below where denitrification is favored, reactions such as Manganese Reduction, Iron Reduction, Sulfate Reduction, Methane Reduction (also known as Methanogenesis), become favored respectively. This favorability is governed by Gibbs Free Energy (ΔG).

Redox Zonation

Redox zonation refers to how the processes that transfer terminal electrons as a result of organic matter degradation vary depending on time and space. Certain reactions will be favored over others due to their energy yield as detailed in the energy acceptor cascade detailed above. In oxic conditions, in which oxygen is readily available, aerobic respiration will be favored due to its high energy yield. Once the use of oxygen through respiration exceeds the input of oxygen due to bioturbation and diffusion, the environment will become anoxic and organic matter will be broken down via other means, such as denitrification and manganese reduction.

Remineralization in the Open Ocean

In most open ocean ecosystems only a small fraction of organic matter reaches the seafloor. Biological activity in the photic zone of most water bodies tends to recycle material so well that only a small fraction of organic matter ever sinks out of that top photosynthetic layer. Remineralization within this top layer occurs rapidly and due to the higher concentrations of organisms and the availability of light, those remineralized nutrients are often taken up by autotrophs just as rapidly as they are released.

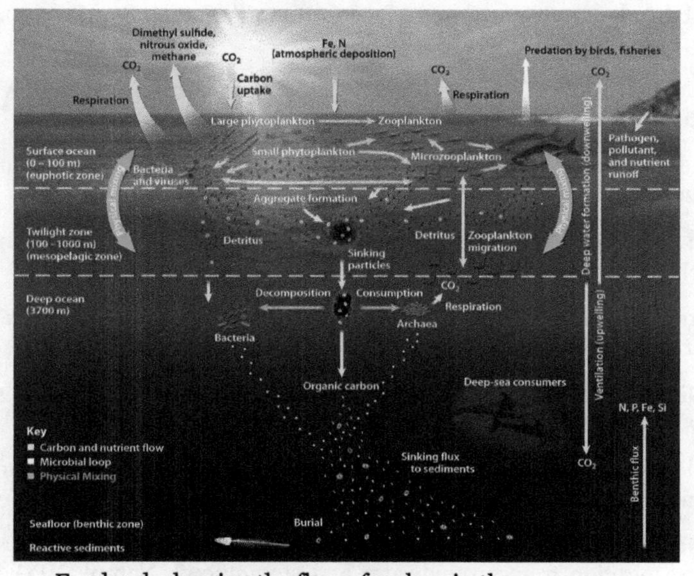

Food web showing the flow of carbon in the open ocean

What fraction does escape varies depending on the location of interest. For example, in the North Sea, values of carbon deposition are ~1% of primary production while that value is <0.5% in the open oceans on average. Therefore, most of nutrients remain in the water column, recycled by the biota. Heterotrophic organisms will utilize the materials produced by the autotrophic (and chemotrophic) organisms and via respiration will remineralize the compounds from the organic form back to inorganic, making them available for primary producers again.

For most areas of the ocean, the highest rates of carbon remineralization occur at depths between 100-1200m in the water column, decreasing down to about 1200m where remineralization rates remain pretty constant at 0.1 μmol kg^{-1} yr^{-1}. As a result of this, the pool of remineralized carbon (which generally takes the form of carbon dioxide) tends to increase in th

Most remineralization is done with dissolved organic carbon (DOC). Studies have shown that it is larger sinking particles that transport matter down to the sea floor while suspended particles and dissolved organics are mostly consumed by remineralization. This happens in part due to the fact that organisms must typically ingest nutrients smaller than they are, often by orders of magnitude. With the microbial community making up 90% of marine biomass, it is particles smaller than the microbes (on the order of 10^{-6}) that will be taken up for remineralization.

References

- Vladimir I. Vernadsky, 2007, Essays on Geochemistry & the Biosphere, tr. Olga Barash, Santa Fe, NM, Synergetic Press, ISBN 0-907791-36-0
- Crowley, T. J. (2000). "Causes of Climate Change Over the Past 1000 Years". Science. 289 (5477): 270–277. Bibcode:2000Sci...289..270C. doi:10.1126/science.289.5477.270. ISSN 0036-8075. PMID 10894770.
- "An Introduction to the Global Carbon Cycle" (PDF). University of New Hampshire. 2009. Retrieved 6 February 2016.
- Planet, The Habitable. "Carbon Cycling and Earth's Climate". Many Planets, One Earth. 4. Retrieved 2012-06-24.
- Buis, Alan; Ramsayer, Kate; Rasmussen, Carol (12 November 2015). "A Breathing Planet, Off Balance". NASA. Retrieved 13 November 2015.
- Staff (12 November 2015). "Audio (66:01) - NASA News Conference - Carbon & Climate Telecon". NASA. Retrieved 12 November 2015.
- St. Fleur, Nicholas (10 November 2015). "Atmospheric Greenhouse Gas Levels Hit Record, Report Says". New York Times. Retrieved 11 November 2015.
- Ritter, Karl (9 November 2015). "UK: In 1st, global temps average could be 1 degree C higher". AP News. Retrieved 11 November 2015.
- Evaristo, Jaivime; Jasechko, Scott; McDonnell, Jeffrey J. (2015-09-03). "Global separation of plant transpiration from groundwater and streamflow". Nature. NPG. 525 (7567): 91–94. Bibcode:2015Natur.525...91E. doi:10.1038/nature14983. ISSN 0028-0836.
- Traufetter, Gerald (January 2, 2009). "Cold Carbon Sink: Slowing Global Warming with Antarctic Iron". Spiegel Online. Retrieved May 9, 2010.

- Jin,X.;Gruber,N.;Frenzel1,H.;Doney,S.C.;McWilliams,J.C.(2008)."TheimpactonatmosphericCO 2 of iron fertilization induced changes in the ocean's biological pump". Biogeosciences. 5: 385–406. doi:10.5194/bg-5-385-2008. Retrieved May 9, 2010.
- Monastersky, Richard (September 30, 1995). "Iron versus the Greenhouse - Oceanographers cautiously explore a global warming therapy". Science News. Retrieved May 9, 2010.
- Monastersky, Richard (September 30, 1995). "Iron versus the Greenhouse: Oceanographers cautiously explore a global warming therapy". Science News. 148 (14): 220. doi:10.2307/4018225. Retrieved August 21, 2015.
- Salleh, Anna (November 9, 2007). "Urea 'climate solution' may backfire". ABC Science. Australian Broadcasting Commission. Retrieved May 9, 2010.
- Pearce, Fred (September 26, 2007). "Ocean pumps could counter global warming". New Scientist. Retrieved May 9, 2010.
- Duke, John H. (2008). "A proposal to force vertical mixing of the Pacific Equatorial Undercurrent to create a system of equatorially trapped coupled convection that counteracts global warming" (PDF). Geophysical Research Abstracts. Retrieved May 9, 2010.
- Flannery, Tim (20 November 2015). "Climate crisis: seaweed, coffee and cement could save the planet". The Guardian. Guardian Media Group. Retrieved 25 November 2015.
- Lovett, Richard (May 3, 2008). "Burying biomass to fight climate change". New Scientist (2654). Retrieved May 9, 2010.
- "International Biochar Initiative | International Biochar Initiative". Biochar-international.org. Retrieved May 9, 2010.

3

Soil Chemistry: A Comprehensive Study

Soil chemistry is the study of the features of the soil. It is mainly affected by mineral composition and environmental factors. The major environmental issues with soil are soil contamination, desertification, land surface effects on climate and soil erosion. Soil chemistry is best understood in confluence with the major topics listed in the following section.

Soil Chemistry

Soil chemistry is the study of the chemical characteristics of soil. Soil chemistry is affected by mineral composition, organic matter and environmental factors.

Soil chemistry is the study of the chemical characteristics of soil. Soil chemistry is affected by mineral composition, organic matter and environmental factors. Soil - (i) The unconsolidated mineral or organic material on the immediate surface of the Earth that serves as a natural medium for the growth of land plants. (ii) The unconsolidated mineral or organic matter on the surface of the Earth that has been subjected to and shows effects of genetic and environmental factors of: climate (including water and temperature effects), and macro- and micro organisms, conditioned by relief, acting on parent material over a period of time. A product-soil differs from the material from which it is derived in many physical, chemical,biological, and morphological properties and characteristics. This definition is from Soil Taxonomy, second edition.

Key facts you should know about Soil:

Soil makes up the outermost layer of our planet and is formed from rocks and decaying plants and Animals. Soil has varying amounts of organic matter (living and dead organisms), minerals, and nutrients. • An average soil sample is 45 percent minerals, 25 percent water, 25 percent air, and five percent organic matter. Different-sized mineral particles, such as sand, silt, and clay, give soil its texture. • Topsoil is the most productive soil layer. • Ten tonnes of topsoil spread evenly over a hectare is only as thick as a one Euro coin. • Natural processes can take more than 500 years to form 2 centimeters of topsoil. • In some cases, 5 tonnes of animal life can live in one hectare of soil. • Fungi and bacteria help break down organic matter in the soil. • Earthworms digest organic matter, recycle nutrients, and make the surface soil richer. • Roots loosen the soil, allowing oxygen to penetrate. This benefits animals living in the soil. They also benefit roots which require oxygen themselves. They further hold soil together and help prevent erosion. • A fully functioning soil reduces the risk of floods and protects underground water supplies by neutralizing or filtering out potential pollutants and storing as much as 3750 tonnes of water per hectare. • Scientists have found 10,000 types of soil in Europe and about 70,000 types of soil in the United States. • Soil stores 10% of the world's carbon dioxide emissions.

Environmental Soil Chemistry

A knowledge of environmental soil chemistry is paramount to predicting the fate of contaminants, as well as the processes by which they are initially released into the soil. Once a chemical is exposed to the soil environment myriad chemical reactions can occur that may increase or decrease contaminant toxicity. These reactions include adsorption/desorption, precipitation, polymerization, dissolution, complexation and oxidation/reduction. These reactions are often disregarded by scientists and engineers involved with environmental remediation. Understanding these processes enable us to better predict the fate and toxicity of contaminants and provide the knowledge to develop scientifically correct, and cost-effective remediation strategies.

Environmental Issues with Soil

Soil Contamination

Excavation showing soil contamination at a disused gasworks.

Soil contamination or soil pollution as part of land degradation is caused by the presence of xenobiotic (human-made) chemicals or other alteration in the natural soil environment. It is typically caused by industrial activity, agricultural chemicals, or improper disposal of waste. The most common chemicals involved are petroleum hydrocarbons, polynuclear aromatic hydrocarbons (such as naphthalene and benzo(a)pyrene), solvents, pesticides, lead, and other heavy metals. Contamination is correlated with the degree of industrialization and intensity of chemical usage.

The concern over soil contamination stems primarily from health risks, from direct contact with the contaminated soil, vapors from the contaminants, and from secondary contamination of water supplies within and underlying the soil. Mapping of contaminated soil sites and the resulting cleanup are time consuming and expensive tasks, requiring extensive amounts of geology, hydrology, chemistry, computer modeling skills, and GIS in Environmental Contamination, as well as an appreciation of the history of industrial chemistry.

In North America and Western Europe the extent of contaminated land is best known, with many of countries in these areas having a legal framework to identify and deal with this environmental problem. Developing countries tend to be less tightly regulated despite some of them having undergone significant industrialization.

Causes

Soil pollution can be caused by the following (non-exhaustive list!):

- Oil drilling
- Mining and activities by other heavy industries
- Accidental spills as may happen during activities, etc.
- Corrosion of underground storage tanks
- Acid rain (in turn caused by air pollution)
- Intensive farming
- Agrochemicals, such as pesticides, herbicides and fertilizers
- Industrial accidents
- Road debris
- Drainage of contaminated surface water into the soil
- Waste disposal
    - Oil and fuel dumping
    - Nuclear wastes
    - Direct discharge of industrial wastes to the soil
    - Landfill and illegal dumping
    - coal ash

- Electronic waste
- ammunitions and agents of war

The most common chemicals involved are petroleum hydrocarbons, solvents, pesticides, lead, and other heavy metals.

In a wider sense, genetically modified plants (GMP) can count as a risk factor for soils, because of their potential to affect the soil fauna. Any activity that leads to other forms of soil degradation (erosion, compaction, etc.) may indirectly worsen the contamination effects in that soil remediation becomes more tedious.

Historical deposition of coal ash used for residential, commercial, and industrial heating, as well as for industrial processes such as ore smelting, were a common source of contamination in areas that were industrialized before about 1960. Coal naturally concentrates lead and zinc during its formation, as well as other heavy metals to a lesser degree. When the coal is burned, most of these metals become concentrated in the ash (the principal exception being mercury). Coal ash and slag may contain sufficient lead to qualify as a "characteristic hazardous waste", defined in the USA as containing more than 5 mg/l of extractable lead using the TCLP procedure. In addition to lead, coal ash typically contains variable but significant concentrations of polynuclear aromatic hydrocarbons (PAHs; e.g., benzo(a)anthracene, benzo(b)fluoranthene, benzo(k)fluoranthene, benzo(a)pyrene, indeno(cd)pyrene, phenanthrene, anthracene, and others). These PAHs are known human carcinogens and the acceptable concentrations of them in soil are typically around 1 mg/kg. Coal ash and slag can be recognised by the presence of off-white grains in soil, gray heterogeneous soil, or (coal slag) bubbly, vesicular pebble-sized grains.

Treated sewage sludge, known in the industry as biosolids, has become controversial as a "fertilizer". As it is the byproduct of sewage treatment, it generally contains more contaminants such as organisms, pesticides, and heavy metals than other soil.

In the European Union, the Urban Waste Water Treatment Directive allows sewage sludge to be sprayed onto land. The volume is expected to double to 185,000 tons of dry solids in 2005. This has good agricultural properties due to the high nitrogen and phosphate content. In 1990/1991, 13% wet weight was sprayed onto 0.13% of the land; however, this is expected to rise 15 fold by 2005. Advocates say there is a need to control this so that pathogenic microorganisms do not get into water courses and to ensure that there is no accumulation of heavy metals in the top soil.

Pesticides and Herbicides

A pesticide is a substance or mixture of substances used to kill a pest. A pesticide may be a chemical substance, biological agent (such as a virus or bacteria), antimicrobial, disinfectant or device used against any pest. Pests include insects, plant pathogens, weeds, mollusks, birds, mammals, fish, nematodes (roundworms) and microbes that compete with humans for food, destroy property, spread or are a vector for disease or cause a nuisance. Although there are benefits to the use of pesticides, there are also drawbacks, such as potential toxicity to humans and other organisms.

Herbicides are used to kill weeds, especially on pavements and railways. They are similar to auxins and most are biodegradable by soil bacteria. However, one group derived from trinitrotoluene (2:4 D and 2:4:5 T) have the impurity dioxin, which is very toxic and causes fatality even in low concentrations. Another herbicide is Paraquat. It is highly toxic but it rapidly degrades in soil due to the action of bacteria and does not kill soil fauna.

Insecticides are used to rid farms of pests which damage crops. The insects damage not only standing crops but also stored ones and in the tropics it is reckoned that one third of the total production is lost during food storage. As with fungicides, the first insecticides used in the nineteenth century were inorganic e.g. Paris Green and other compounds of arsenic. Nicotine has also been used since the late eighteenth century.

There are now two main groups of synthetic insecticides -

1. Organochlorines include DDT, Aldrin, Dieldrin and BHC. They are cheap to produce, potent and persistent. DDT was used on a massive scale from the 1930s, with a peak of 72,000 tonnes used 1970. Then usage fell as the harmful environmental effects were realized. It was found worldwide in fish and birds and was even discovered in the snow in the Antarctic. It is only slightly soluble in water but is very soluble in the bloodstream. It affects the nervous and endocrine systems and causes the eggshells of birds to lack calcium causing them to be easily breakable. It is thought to be responsible for the decline of the numbers of birds of prey like ospreys and peregrine falcons in the 1950s - they are now recovering. As well as increased concentration via the food chain, it is known to enter via permeable membranes, so fish get it through their gills. As it has low water solubility, it tends to stay at the water surface, so organisms that live there are most affected. DDT found in fish that formed part of the human food chain caused concern, but the levels found in the liver, kidney and brain tissues was less than 1 ppm and in fat was 10 ppm, which was below the level likely to cause harm. However, DDT was banned in the UK and the United States to stop the further buildup of it in the food chain. U.S. manufacturers continued to sell DDT to developing countries, who could not afford the expensive replacement chemicals and who did not have such stringent regulations governing the use of pesticides..

2. Organophosphates, e.g. parathion, methyl parathion and about 40 other insecticides are available nationally. Parathion is highly toxic, methyl-parathion is less so and Malathion is generally considered safe as it has low toxicity and is rapidly broken down in the mammalian liver. This group works by preventing normal nerve transmission as cholinesterase is prevented from breaking down the transmitter substance acetylcholine, resulting in uncontrolled muscle movements.

Agents of War

The disposal of munitions, and a lack of care in manufacture of munitions caused by the urgency of production, can contaminate soil for extended periods. There is little published evidence on this type of contamination largely because of restrictions placed by Governments of many countries on the publication of material related to war effort. However, mustard gas stored during World War II has contaminated some sites for up to 50 years and the testing of Anthrax as a potential biological weapon contaminated the whole island of Gruinard

Health Effects

Contaminated or polluted soil directly affects human health through direct contact with soil or via inhalation of soil contaminants which have vaporized; potentially greater threats are posed by the infiltration of soil contamination into groundwater aquifers used for human consumption, sometimes in areas apparently far removed from any apparent source of above ground contamination. This tends to result in the development of pollution-related diseases.

Health consequences from exposure to soil contamination very greatly depending on pollutant type, pathway of attack and vulnerability of the exposed population. Chronic exposure to chromium, lead and other metals, petroleum, solvents, and many pesticide and herbicide formulations can be carcinogenic, can cause congenital disorders, or can cause other chronic health conditions. Industrial or man-made concentrations of naturally occurring substances, such as nitrate and ammonia associated with livestock manure from agricultural operations, have also been identified as health hazards in soil and groundwater.

Chronic exposure to benzene at sufficient concentrations is known to be associated with higher incidence of leukemia. Mercury and cyclodienes are known to induce higher incidences of kidney damage and some irreversible diseases. PCBs and cyclodienes are linked to liver toxicity. Organophosphates and carbonates can induce a chain of responses leading to neuromuscular blockage. Many chlorinated solvents induce liver changes, kidney changes and depression of the central nervous system. There is an entire spectrum of further health effects such as headache, nausea, fatigue, eye irritation and skin rash for the above cited and other chemicals. At sufficient dosages a large number of soil contaminants can cause death by exposure via direct contact, inhalation or ingestion of contaminants in groundwater contaminated through soil.

The Scottish Government has commissioned the Institute of Occupational Medicine to undertake a review of methods to assess risk to human health from contaminated land. The overall aim of the project is to work up guidance that should be useful to Scottish Local Authorities in assessing whether sites represent a significant possibility of significant harm (SPOSH) to human health. It is envisaged that the output of the project will be a short document providing high level guidance on health risk assessment with reference to existing published guidance and methodologies that have been identified as being particularly relevant and helpful. The project will examine how policy guidelines have been developed for determining the acceptability of risks to human health and propose an approach for assessing what constitutes unacceptable risk in line with the criteria for SPOSH as defined in the legislation and the Scottish Statutory Guidance.

Ecosystem Effects

Not unexpectedly, soil contaminants can have significant deleterious consequences for ecosystems. There are radical soil chemistry changes which can arise from the presence of many hazardous chemicals even at low concentration of the contaminant species. These changes can manifest in the alteration of metabolism of endemic microorganisms and arthropods resident in a given soil environment. The result can be virtual eradication of some of the primary food chain, which in turn could have major consequences for predator or consumer species. Even if the chemical effect on lower life forms is small, the lower pyramid levels of the food chain may ingest alien chemicals, which normally become more concentrated for each consuming rung of the food chain. Many of these effects are now well known, such as the concentration of persistent DDT materials for avian consumers, leading to weakening of egg shells, increased chick mortality and potential extinction of species.

Effects occur to agricultural lands which have certain types of soil contamination. Contaminants typically alter plant metabolism, often causing a reduction in crop yields. This has a secondary effect upon soil conservation, since the languishing crops cannot shield the Earth's soil from erosion. Some of these chemical contaminants have long half-lives and in other cases derivative chemicals are formed from decay of primary soil contaminants.

Cleanup Options

Cleanup or environmental remediation is analyzed by environmental scientists who utilize field measurement of soil chemicals and also apply computer models (GIS in Environmental Contamination) for analyzing transport and fate of soil chemicals. Various technologies have been developed for remediation of oil-contaminated soil/ sediments There are several principal strategies for remediation:

- Excavate soil and take it to a disposal site away from ready pathways for human or sensitive ecosystem contact. This technique also applies to dredging of bay muds containing toxins.
- Aeration of soils at the contaminated site (with attendant risk of creating air pollution)
- Thermal remediation by introduction of heat to raise subsurface temperatures sufficiently high to volatize chemical contaminants out of the soil for vapour

extraction. Technologies include ISTD, electrical resistance heating (ERH), and ET-DSPtm.

- Bioremediation, involving microbial digestion of certain organic chemicals. Techniques used in bioremediation include landfarming, biostimulation and bioaugmentating soil biota with commercially available microflora.
- Extraction of groundwater or soil vapor with an active electromechanical system, with subsequent stripping of the contaminants from the extract.
- Containment of the soil contaminants (such as by capping or paving over in place).
- Phytoremediation, or using plants (such as willow) to extract heavy metals.
- Mycoremediation, or using fungus to metabolize contaminants and accumulate heavy metals.
- Remediation of oil contaminated sediments with self-collapsing air microbubbles.

By Country

Various national standards for concentrations of particular contaminants include the United States EPA Region 9 Preliminary Remediation Goals (U.S. PRGs), the U.S. EPA Region 3 Risk Based Concentrations (U.S. EPA RBCs) and National Environment Protection Council of Australia Guideline on Investigation Levels in Soil and Groundwater.

People's Republic of China

The immense and sustained growth of the People's Republic of China since the 1970s has exacted a price from the land in increased soil pollution. The State Environmental Protection Administration believes it to be a threat to the environment, to food safety and to sustainable agriculture. According to a scientific sampling, 150 million mi (100,000 square kilometers) of China's cultivated land have been polluted, with contaminated water being used to irrigate a further 31.5 million mi (21,670 km^2.) and another 2 million mi (1,300 square kilometers) covered or destroyed by solid waste. In total, the area accounts for one-tenth of China's cultivatable land, and is mostly in economically developed areas. An estimated 12 million tonnes of grain are contaminated by heavy metals every year, causing direct losses of 20 billion yuan (US$2.57 billion).

United Kingdom

Generic guidance commonly used in the United Kingdom are the Soil Guideline Values published by the Department for Environment, Food and Rural Affairs (DEFRA) and the Environment Agency. These are screening values that demonstrate the minimal acceptable level of a substance. Above this there can be no assurances in terms of significant risk of harm to human health. These have been derived using the Contaminated Land Exposure Assessment Model (CLEA UK). Certain input parameters such as Health Criteria Values, age and land use are fed into CLEA UK to obtain a probabilistic output.

Guidance by the Inter Departmental Committee for the Redevelopment of Contaminated Land (ICRCL) has been formally withdrawn by DEFRA, for use as a prescriptive document to determine the potential need for remediation or further assessment.

The CLEA model published by DEFRA and the Environment Agency (EA) in March 2002 sets a framework for the appropriate assessment of risks to human health from contaminated land, as required by Part IIA of the Environmental Protection Act 1990. As part of this framework, generic Soil Guideline Values (SGVs) have currently been derived for ten contaminants to be used as "intervention values". These values should not be considered as remedial targets but values above which further detailed assessment should be considered;

Three sets of CLEA SGVs have been produced for three different land uses, namely

- residential (with and without plant uptake)
- allotments
- commercial/industrial

It is intended that the SGVs replace the former ICRCL values. It should be noted that the CLEA SGVs relate to assessing chronic (long term) risks to human health and do not apply to the protection of ground workers during construction, or other potential receptors such as groundwater, buildings, plants or other ecosystems. The CLEA SGVs are not directly applicable to a site completely covered in hardstanding, as there is no direct exposure route to contaminated soils.

To date, the first ten of fifty-five contaminant SGVs have been published, for the following: arsenic, cadmium, chromium, lead, inorganic mercury, nickel, selenium ethyl benzene, phenol and toluene. Draft SGVs for benzene, naphthalene and xylene have been produced but their publication is on hold. Toxicological data (Tox) has been published for each of these contaminants as well as for benzo[a]pyrene, benzene, dioxins, furans and dioxin-like PCBs, naphthalene, vinyl chloride, 1,1,2,2 tetrachloroethane and 1,1,1,2 tetrachloroethane, 1,1,1 trichloroethane, tetrachloroethene, carbon tetrachloride, 1,2-dichloroethane, trichloroethene and xylene. The SGVs for ethyl benzene, phenol and toluene are dependent on the soil organic matter (SOM) content (which can be calculated from the total organic carbon (TOC) content). As an initial screen the SGVs for 1% SOM are considered to be appropriate.

India

In March 2009, the issue of Uranium poisoning in Punjab attracted press coverage. It was alleged to be caused by fly ash ponds of thermal power stations, which reportedly lead to severe birth defects in children in the Faridkot and Bhatinda districts of Punjab. The news reports claimed the uranium levels were more than 60 times the maximum safe limit. In 2012, the Government of India confirmed that the ground water in Malwa belt of Punjab has uranium metal that is 50% above the trace limits set by the United Nations' World Health Organization. Scientific studies, based on over 1000 samples from various sampling points, could not trace the source to fly ash and any sources from thermal power plants or industry as originally alleged. The study also revealed that the uranium concentration in ground water of Malwa district is not 60 times the WHO limits, but only 50% above the WHO limit in 3 locations. This highest concentration found in samples was less than those found naturally in ground waters currently used for human purposes elsewhere, such as Finland. Research is underway to identify natural or other sources for the uranium.

Desertification

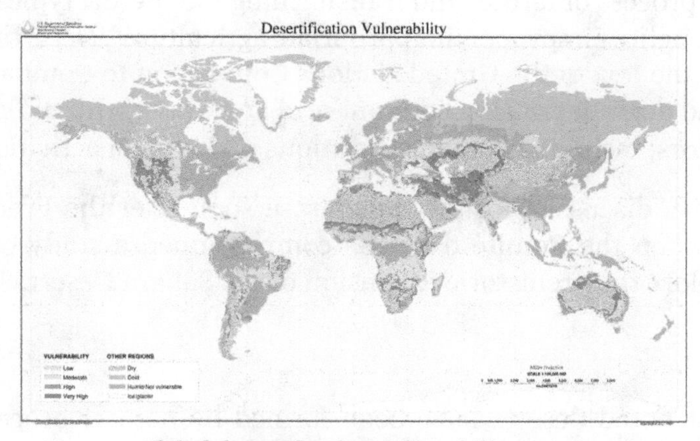

Global desertification vulnerability map

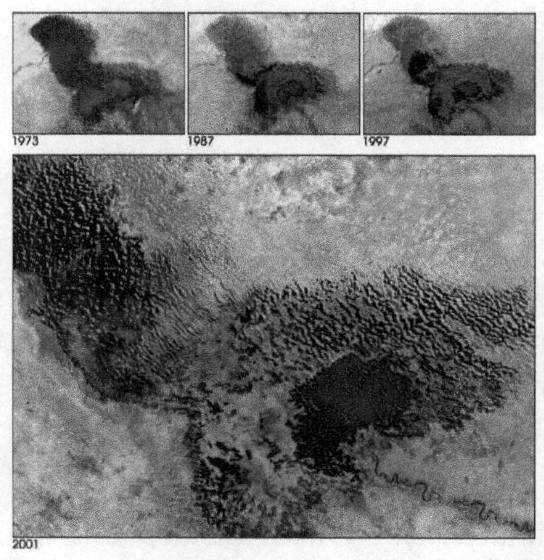

Lake Chad in a 2001 satellite image, with the actual lake in blue. The lake has shrunk by 94% since the 1960s.

Desertification is a type of land degradation in which relatively dry area of land becomes increasingly arid, typically losing its bodies of water as well as vegetation and wildlife. It is caused by a variety of factors, such as through climate change and through the overexploitation of soil through humankind's undertaking. When deserts appear automatically over the natural course of a planet's life cycle, then it can be called a natural phenomenon; however, when deserts emerge due to the rampant and unchecked depletion of nutrients in soil that are essential for it to remain airable, then a virtual "soil death" can be spoken of, which traces its cause back to human overexploitation. Desertification is a significant global ecological and environmental problem.

Definitions

Considerable controversy exists over the proper definition of the term "desertification" for which Helmut Geist (2005) has identified more than 100 formal definitions. The most widely accepted of these is that of the Princeton University Dictionary which defines it as "the process of fertile land transforming into desert typically as a result of deforestation, drought or improper/inappropriate agriculture". Desertification has been neatly defined in the text of the United Nations Convention to Combat Desertification (UNCCD) as "land degradation in arid, semi-arid and dry sub-humid regions resulting from various factors, including climatic variations and human activities."

The earliest known discussion of the topic arose soon after the French colonization of West Africa, when the Comité d'Etudes commissioned a study on *desséchement progressif* to explore the prehistoric expansion of the Sahara Desert.

History

The world's most noted deserts have been formed by natural processes interacting over long intervals of time. During most of these times, deserts have grown and shrunk independent of human activities. Paleodeserts are large sand seas now inactive because they are stabilized by vegetation, some extending beyond the present margins of core deserts, such as the Sahara, the largest hot desert.

Desertification has played a significant role in human history, contributing to the collapse of several large empires, such as Carthage, Greece, and the Roman Empire, as well as causing displacement of local populations. Historical evidence shows that the serious and extensive land deterioration occurring several centuries ago in arid regions had three epicenters: the Mediterranean, the Mesopotamian Valley, and the Loess Plateau of China, where population was dense.

Areas Affected

Drylands occupy approximately 40–41% of Earth's land area and are home to more than 2 billion people. It has been estimated that some 10–20% of drylands are already degraded, the total area affected by desertification being between 6 and 12 million square kilometres, that about 1–6% of the inhabitants of drylands live in desertified areas, and that a billion people are under threat from further desertification.

Sun, Moon, and large telescopes above the Desert

As of 1998, the then-current degree of southward expansion of the Sahara was not well known, due to a lack of recent, measurable expansion of the desert into the Sahel at the time.

Causes of desertification in Sahel:

The impact of global warming and human activities are presented in the Sahel. In this area, the level of desertification is very high compared to other areas in the world.

All areas situated in the eastern part of Africa (i.e. in the Sahel region) are characterized by a dry climate, hot temperatures, and low rainfall (300–750 mm rainfall per year). So, droughts are the rule in the Sahel region.

Development of the Desertification Process in Sahel:

Some studies have shown that Africa has lost approximately 650 000 km^2 of its productive agricultural land over the past 50 years. The propagation of desertification in this area is considerable.

Some statistics have proved that since 1900, the Sahara has expanded by 250 km, covering an additional area of 6000 square kilometers.

Impacts of Desertification in Sahel:

The survey, done by the research institute for development, had demonstrated that this dryness is spreading fast in the Sahelian countries. desertification in the Sahel can affect more than one billion of its inhabitants. 70% of the arid area has deteriorated and water resources have disappeared, leading to soil degradation. The loss of topsoil means that plants cannot take root firmly and can be uprooted by torrential water or strong winds.

The United Nations Convention (UNC) says that about six million Sahelian citizens would have to give up the desertified zones of sub-Saharan Africa for North Africa and Europe between 1997 and 2020.

Vegetation Patterning

As the desertification takes place, the landscape may progress through different stages and continuously transform in appearance. On gradually sloped terrain, desertification can create increasingly larger empty spaces over a large strip of land, a phenomenon known as “Brousse tigrée”. A mathematical model of this phenomenon proposed by C. Klausmeier attributes this patterning to dynamics in plant-water interaction. One outcome of this observation suggests an optimal planting strategy for agriculture in arid environments.

Causes

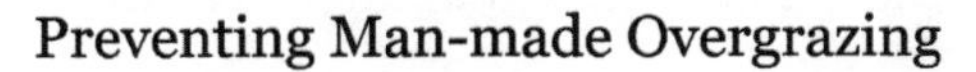
Preventing Man-made Overgrazing

Goats inside of a pen in Norte Chico, Chile. Overgrazing of drylands by poorly managed traditional herding is one of the primary causes of desertification.

Wildebeest in Masai Mara during the Great Migration. Overgrazing is not caused by nomadic grazers in huge populations of travel herds, nor by holistic planned grazing.

The immediate cause is the loss of most vegetation. This is driven by a number of factors, alone or in combination, such as drought, climatic shifts, tillage for agriculture, overgrazing and deforestation for fuel or construction materials.

Vegetation plays a major role in determining the biological composition of the soil. Studies have shown that, in many environments, the rate of erosion and runoff decreases exponentially with increased vegetation cover. Unprotected, dry soil surfaces blow away with the wind or are washed away by flash floods, leaving infertile lower soil layers that bake in the sun and become an unproductive hardpan. Controversially, Allan Savory has claimed that the controlled movement of herds of livestock, mimicking herds of grazing wildlife, can reverse desertification.

A shepherd guiding his sheep through the high desert outside of Marrakech, Morocco

Poverty

At least 90% of the inhabitants of drylands live in developing nations, where they also suffer from poor economic and social conditions. This situation is exacerbated by land degradation because of the reduction in productivity, the precariousness of living conditions and the difficulty of access to resources and opportunities.

A downward spiral is created in many underdeveloped countries by overgrazing, land exhaustion and overdrafting of groundwater in many of the marginally productive world regions due to overpopulation pressures to exploit marginal drylands for farming. Decision-makers are understandably averse to invest in arid zones with low potential. This absence of investment contributes to the marginalisation of these zones. When unfavourable agro-climatic conditions are combined with an absence of infrastructure and access to markets, as well as poorly adapted production techniques and an underfed and undereducated population, most such zones are excluded from development.

Desertification often causes rural lands to become unable to support the same sized populations that previously lived there. This results in mass migrations out of rural areas and into urban areas, particularly in Africa. These migrations into the cities often cause large numbers of unemployed people, who end up living in slums.

Countermeasures and Prevention

Techniques and countermeasures exist for mitigating or reversing the effects of desertification, and some possess varying levels of difficulty. For some, there are numerous barriers to their implementation. Yet for others, the solution simply requires the exercise of human reason.

Anti-sand shields in north Sahara, Tunisia

Jojoba plantations, such as those shown, have played a role in combating edge effects of desertification in the Thar Desert, India.

One less difficult solution that has been proposed, however controversial it may be, is to bring about a cap on the population growth, and in fact to turn this into a population decay, so that each year there will gradually exist fewer and fewer humans who require the land to be depleted even further in order to grow their food.

Proponents of this solution claim that this would put the economy into dire straits, however many unemployed people (estimated at approximately 592,398,142 when the average unemployment rate in 2016 (8 percent) is applied to the recorded global population of 7,404,976,783) already exist who could rise up to the duty of filling jobs.

One proposed barrier is that the costs of adopting sustainable agricultural practices sometimes exceed the benefits for individual farmers, even while they are socially and environmentally beneficial. Another issue is a lack of political will, and lack of funding to support land reclamation and anti-desertification programs.

Desertification is recognized as a major threat to biodiversity. Some countries have developed Biodiversity Action Plans to counter its effects, particularly in relation to the protection of endangered flora and fauna.

Reforestation gets at one of the root causes of desertification and is not just a treatment of the symptoms. Environmental organizations work in places where deforestation and desertification are contributing to extreme poverty. There they focus primarily on educating the local population about the dangers of deforestation and sometimes employ them to grow seedlings, which they transfer to severely deforested areas during the rainy season. The Food and Agriculture Organization of the United Nations launched the FAO Drylands Restoration Initiative in 2012 to draw together knowledge and experience on dryland restoration. In 2015, FAO published global guidelines for the restoration of degraded forests and landscapes in drylands, in collaboration with the Turkish Ministry of Forestry and Water Affairs and the Turkish Cooperation and Coordination Agency.

Techniques focus on two aspects: provisioning of water, and fixation and hyper-fertilizing soil.

Fixating the soil is often done through the use of shelter belts, woodlots and windbreaks. Windbreaks are made from trees and bushes and are used to reduce soil erosion and evapotranspiration. They were widely encouraged by development agencies from the middle of the 1980s in the Sahel area of Africa.

Some soils (for example, clay), due to lack of water can become consolidated rather than porous (as in the case of sandy soils). Some techniques as zaï or tillage are then used to still allow the planting of crops.

Another technique that is useful is contour trenching. This involves the digging of 150m long, 1m deep trenches in the soil. The trenches are made parallel to the height lines of the landscape, preventing the water from flowing within the trenches and causing erosion. Stone walls are placed around the trenches to prevent the trenches from closing up again. The method was invented by Peter Westerveld.

Enriching of the soil and restoration of its fertility is often done by plants. Of these, the Leguminous plants which extract nitrogen from the air and fixes it in the soil, and food crops/trees as grains, barley, beans and dates are the most important. Sand fences can also be used to control drifting of soil and sand erosion.

Some research centra (such as Bel-Air Research Center IRD/ISRA/UCAD) are also experimenting with the inoculation of tree species with Mycorrhiza in arid zones. The mycorrhiza are basically fungi attaching themselves to the roots of the plants. They hereby create a symbiotic relation with the trees, increasing the surface area of the tree's roots greatly (allowing the tree to gather much more nutrients from the soil). An example of a promosing setup is Jujube combined with Glomus aggregatum.

As there are many different types of deserts, there are also different types of desert reclamation methodologies. An example for this is the salt-flats in the Rub' al Khali desert in Saudi-Arabia. These salt-flats are one of the most promising desert areas for seawater agriculture and could be revitalized without the use of freshwater or much energy.

Farmer-managed natural regeneration (FMNR) is another technique that has produced successful results for desert reclamation. Since 1980, this method to reforest degraded landscape has been applied with some success in Niger. This simple and low-cost method has enabled farmers to regenerate some 30,000 square kilometers in Niger. The process involves enabling native sprouting tree growth through selective pruning of shrub shoots. The residue from pruned trees can be used to provide mulching for fields thus increasing soil water retention and reducing evaporation. Additionally, properly spaced and pruned trees can increase crop yields. The Humbo Assisted Regeneration Project which uses FMNR techniques in Ethiopia has received money from The World Bank's BioCarbon Fund, which supports projects that sequester or conserve carbon in forests or agricultural ecosystems.

Managed grazing methods are argued to be able to restore grasslands.

Managed Grazing

Restoring grasslands store CO_2 from the air into plant material. Grazing livestock, usually not left to wander, would eat the grass and would minimize any grass growth while grass left alone would eventually grow to cover its own growing buds, preventing them from photosynthesizing and killing the plant. A method proposed to restore grasslands uses fences with many small paddocks and moving herds from one paddock to another after a day or two in order to mimic natural grazers and allowing the grass to grow optimally. It is estimated that increasing the carbon content of the soils in the world's 3.5 billion hectares of agricultural grassland would offset nearly 12 years of CO_2 emissions. Allan Savory, as part of holistic management, claims that while large herds are often blamed for desertification, prehistoric lands used to support large or larger herds and areas where herds were removed in the United States are still desertifying.

Land Surface Effects on Climate

Land surface effects on climate are wide-ranging and vary by region. Deforestation and exploitation of natural landscapes play a significant role. Some of these environmental changes are similar to those caused by the effects of global warming.

Deforestation Effects

Major land surface changes affecting climate include deforestation (especially in tropical areas), and destruction of grasslands and xeric woodlands by overgrazing, or lack of grazing. These changes in the natural landscape reduce evapotranspiration, and thus water vapor, in the atmosphere, limiting clouds and precipitation (this may contribute to the retreat of glaciers). It has been proposed, in the journal Atmospheric Chemistry and Physics, that evaporation rates from forested areas may exceed that of the oceans, creating zones of low pressure, which enhance the development of storms and rainfall through atmospheric moisture recycling. The American Institute of Biological Sciences published a similar paper in support of this concept in 2009. In addition, with deforestation and/or destruction of grasslands, the amount of dew harvested (or condensed) by plants is greatly diminished. All of this helps lead to desertification in these regions.

This concept of land-atmosphere feedback is common among permaculturists, such as Masanobu Fukuoka, who, in his book, *The One Straw Revolution*, said "rain comes from the ground, not the sky."

Deforestation, and conversion of grasslands to desert, may also lead to cooling of the regional climate. This is because of the albedo effect (sunlight reflected by bare ground) during the day, and rapid radiation of heat into space at night, due to the lack of vegetation and atmospheric moisture.

Reforestation, conservation grazing, holistic land management, and, in drylands, water harvesting and keyline design, are examples of methods that might help prevent or lessen these drying effects.

Mountain Meteorological Effects

Orographic Lift

Orographic lift occurs when an air mass is forced from a low elevation to a higher elevation as it moves over rising terrain. As the air mass gains altitude it quickly cools down adiabatically, which can raise the relative humidity to 100% and create clouds and, under the right conditions, precipitation.

Rain Shadow

A rain shadow is a dry area on the leeward side of a mountainous area (away from the wind). The mountains block the passage of rain-producing weather systems and cast a "shadow" of dryness behind them. Wind and moist air is drawn by the prevailing winds towards the top of the mountains, where it condenses and precipitates before it crosses the top. In an effect opposite that of orographic lift, the air, without much moisture left, advances behind the mountains creating a drier side called the "rain shadow".

Foehn Wind

Föhn can be initiated when deep low pressures move into Europe drawing moist Mediterranean air over the Alps.

A föhn or foehn is a type of dry, warm, down-slope wind that occurs in the lee (downwind side) of a mountain range.

It is a rain shadow wind that results from the subsequent adiabatic warming of air that has dropped most of its moisture on windward slopes. As a consequence of the different adiabatic lapse rates of moist and dry air, the air on the leeward slopes becomes warmer than equivalent elevations on the windward slopes. Föhn winds can raise temperatures by as much as 14 °C (25 °F) in just a matter of minutes. Central Europe enjoys a warmer climate due to the Föhn, as moist winds off the Mediterranean Sea blow over the Alps.

Soil Erosion

An actively eroding rill on an intensively-farmed field in eastern Germany

Soil erosion is one form of soil degradation. The erosion of soil is a naturally occurring process on all land. The agents of soil erosion are water and wind, each contributing a significant amount of soil loss each year. Soil erosion may be a slow process that continues relatively unnoticed, or it may occur at an alarming rate causing serious loss of topsoil. The loss of soil from farmland may be reflected in reduced crop production potential, lower surface water quality and damaged drainage networks.

While erosion is a natural process, human activities have increased by 10-40 times the rate at which erosion is occurring globally. Excessive (or accelerated) erosion causes both "on-site" and "off-site" problems. On-site impacts include decreases in agricultural productivity and (on natural landscapes) ecological collapse, both because of loss of the nutrient-rich upper soil layers. In some cases, the eventual end result is desertification. Off-site effects include sedimentation of waterways and eutrophication of water bodies, as well as sediment-related damage to roads and houses. Water and wind erosion are the two primary causes of land degradation; combined, they are responsible for about 84% of the global extent of degraded land, making excessive erosion one of the most significant environmental problems world-wide.

Intensive agriculture, deforestation, roads, anthropogenic climate change and urban sprawl are amongst the most significant human activities in regard to their effect on stimulating erosion. However, there are many prevention and remediation practices that can curtail or limit erosion of vulnerable soils.

Physical Processes

Rainfall and Surface Runoff

Soil and water being splashed by the impact of a single raindrop.

Rainfall, and the surface runoff which may result from rainfall, produces four main types of soil erosion: *splash erosion*, *sheet erosion*, *rill erosion*, and *gully erosion*. Splash erosion is generally seen as the first and least severe stage in the soil erosion process, which is followed by sheet erosion, then rill erosion and finally gully erosion (the most severe of the four).

In *splash erosion*, the impact of a falling raindrop creates a small crater in the soil, ejecting soil particles. The distance these soil particles travel can be as much as 0.6 m (two feet) vertically and 1.5 m (five feet) horizontally on level ground.

If the soil is saturated, or if the rainfall rate is greater than the rate at which water can infiltrate into the soil, surface runoff occurs. If the runoff has sufficient flow energy, it will transport loosened soil particles (sediment) down the slope. *Sheet erosion* is the transport of loosened soil particles by overland flow.

A spoil tip covered in rills and gullies due to erosion processes caused by rainfall: Rummu, Estonia

Rill erosion refers to the development of small, ephemeral concentrated flow paths which function as both sediment source and sediment delivery systems for erosion on hillslopes. Generally, where water erosion rates on disturbed upland areas are greatest, rills are active. Flow depths in rills are typically of the order of a few centimeters (about an inch) or less and along-channel slopes may be quite steep. This means that rills exhibit hydraulic physics very different from water flowing through the deeper, wider channels of streams and rivers.

Gully erosion occurs when runoff water accumulates and rapidly flows in narrow channels during or immediately after heavy rains or melting snow, removing soil to a considerable depth.

Rivers and Streams

Dobbingstone Burn, Scotland—This photo illustrates two different types of erosion affecting the same place. Valley erosion is occurring due to the flow of the stream, and the boulders and stones (and much of the soil) that are lying on the edges are glacial till that was left behind as ice age glaciers flowed over the terrain.

Valley or *stream erosion* occurs with continued water flow along a linear feature. The erosion is both downward, deepening the valley, and headward, extending the valley into the hillside, creating head cuts and steep banks. In the earliest stage of stream erosion, the erosive activity is dominantly vertical, the valleys have a typical **V** cross-section and the stream gradient is relatively steep. When some base level is reached, the erosive activity switches to lateral erosion, which widens the valley floor and creates a narrow floodplain. The stream gradient becomes nearly flat, and lateral deposition of sediments becomes important as the stream meanders across the valley floor. In all stages of stream erosion, by far the most erosion occurs during times of flood, when more and faster-moving water is available to carry a larger sediment load. In such processes, it is not the water alone that erodes: suspended abrasive particles, pebbles and boulders can also act erosively as they traverse a surface, in a process known as *traction*.

Bank erosion is the wearing away of the banks of a stream or river. This is distinguished from changes on the bed of the watercourse, which is referred to as *scour*. Erosion and changes in the form of river banks may be measured by inserting metal rods into the bank and marking the position of the bank surface along the rods at different times.

Thermal erosion is the result of melting and weakening permafrost due to moving water. It can occur both along rivers and at the coast. Rapid river channel migration observed in the Lena River of Siberia is due to thermal erosion, as these portions of the banks are composed of permafrost-cemented non-cohesive materials. Much of this erosion occurs as the weakened banks fail in large slumps. Thermal erosion also affects the Arctic coast, where wave action and near-shore temperatures combine to undercut permafrost bluffs along the shoreline and cause them to fail. Annual erosion rates along a 100-kilometre (62-mile) segment of the Beaufort Sea shoreline averaged 5.6 metres (18 feet) per year from 1955 to 2002.

Floods

At extremely high flows, kolks, or vortices are formed by large volumes of rapidly rushing water. Kolks cause extreme local erosion, plucking bedrock and creating pothole-type geographical features called Rock-cut basins. Examples can be seen in the flood regions result from glacial Lake Missoula, which created the channeled scablands in the Columbia Basin region of eastern Washington.

Wind Erosion

Wind erosion is a major geomorphological force, especially in arid and semi-arid regions. It is also a major source of land degradation, evaporation, desertification, harmful airborne dust, and crop damage—especially after being increased far above natural rates by human activities such as deforestation, urbanization, and agriculture.

Árbol de Piedra, a rock formation in the Altiplano, Bolivia sculpted by wind erosion.

Wind erosion is of two primary varieties: *deflation*, where the wind picks up and carries away loose particles; and *abrasion*, where surfaces are worn down as they are struck by airborne particles carried by wind. Deflation is divided into three categories: (1) *surface creep*, where larger, heavier particles slide or roll along the ground; (2) *saltation*, where particles are lifted a short height into the air, and bounce and saltate across the surface of the soil; and (3) *suspension*, where very small and light particles are lifted into the air by the wind, and are often carried for long distances. Saltation is responsible for the majority (50-70%) of wind erosion, followed by suspension (30-40%), and then surface creep (5-25%).

Wind erosion is much more severe in arid areas and during times of drought. For example, in the Great Plains, it is estimated that soil loss due to wind erosion can be as much as 6100 times greater in drought years than in wet years.

Mass Movement

Mass movement is the downward and outward movement of rock and sediments on a sloped surface, mainly due to the force of gravity.

Wadi in Makhtesh Ramon, Israel, showing gravity collapse erosion on its banks.

Mass movement is an important part of the erosional process, and is often the first stage in the breakdown and transport of weathered materials in mountainous areas. It moves material from higher elevations to lower elevations where other eroding agents such as streams and glaciers can then pick up the material and move it to even lower elevations. Mass-movement processes are always occurring continuously on all slopes; some mass-movement processes act very slowly; others occur very suddenly, often with disastrous results. Any perceptible down-slope movement of rock or sediment is often referred to in general terms as a landslide. However, landslides can be classified in a much more detailed way that reflects the mechanisms responsible for the movement and the velocity at which the movement occurs. One of the visible topographical manifestations of a very slow form of such activity is a scree slope.

Slumping happens on steep hillsides, occurring along distinct fracture zones, often within materials like clay that, once released, may move quite rapidly downhill. They will often show a spoon-shaped isostatic depression, in which the material has begun to slide downhill. In some cases, the slump is caused by water beneath the slope weakening it. In many cases it is simply the result of poor engineering along highways where it is a regular occurrence.

Surface creep is the slow movement of soil and rock debris by gravity which is usually not perceptible except through extended observation. However, the term can also describe the rolling of dislodged soil particles 0.5 to 1.0 mm (0.02 to 0.04 in) in diameter by wind along the soil surface.

Factors Affecting Soil Erosion

Climate

The amount and intensity of precipitation is the main climatic factor governing soil erosion by water. The relationship is particularly strong if heavy rainfall occurs at times when, or in locations where, the soil's surface is not well protected by vegetation. This might be during periods when agricultural activities leave the soil bare, or in semi-arid regions where vegetation is naturally sparse. Wind erosion requires strong winds, particularly during times of drought when vegetation is sparse and soil is dry (and so is more erodible). Other climatic factors such as average temperature and temperature range may also affect erosion, via their effects on vegetation and soil properties. In general, given similar vegetation and ecosystems, areas with more precipitation (especially high-intensity rainfall), more wind, or more storms are expected to have more erosion.

In some areas of the world (e.g. the mid-western USA), rainfall intensity is the primary determinant of erosivity, with higher intensity rainfall generally resulting in more soil erosion by water. The size and velocity of rain drops is also an important factor. Larger and higher-velocity rain drops have greater kinetic energy, and thus their impact will displace soil particles by larger distances than smaller, slower-moving rain drops.

In other regions of the world (e.g. western Europe), runoff and erosion result from relatively low intensities of stratiform rainfall falling onto previously saturated soil. In such situations, rainfall amount rather than intensity is the main factor determining the severity of soil erosion by water.

Soil Structure and Composition

Erosional gully in unconsolidated Dead Sea (Israel) sediments along the southwestern shore. This gully was excavated by floods from the Judean Mountains in less than a year.

The composition, moisture, and compaction of soil are all major factors in determining the erosivity of rainfall. Sediments containing more clay tend to be more resistant to erosion than those with sand or silt, because the clay helps bind soil particles together. Soil containing high levels of organic materials are often more resistant to erosion, because the organic materials coagulate soil colloids and create a stronger, more stable soil structure. The amount of water present in the soil before the precipitation also plays an important role, because it sets limits on the amount of water that can be absorbed by the soil (and hence prevented from flowing on the surface as erosive runoff). Wet, saturated soils will not be able to absorb as much rain water, leading to higher levels of surface runoff and thus higher erosivity for a given volume of rainfall. Soil compaction also affects the permeability of the soil to water, and hence the amount of water that flows away as runoff. More compacted soils will have a larger amount of surface runoff than less compacted soils.

Vegetative Cover

Vegetation acts as an interface between the atmosphere and the soil. It increases the permeability of the soil to rainwater, thus decreasing runoff. It shelters the soil from winds, which results in decreased wind erosion, as well as advantageous changes in microclimate. The roots of the plants bind the soil together, and interweave with other roots, forming a more solid mass that is less susceptible to both water and wind erosion. The removal of vegetation increases the rate of surface erosion.

Topography

The topography of the land determines the velocity at which surface runoff will flow, which in turn determines the erosivity of the runoff. Longer, steeper slopes (especially those without adequate vegetative cover) are more susceptible to very high rates of erosion during heavy rains than shorter, less steep slopes. Steeper terrain is also more prone to mudslides, landslides, and other forms of gravitational erosion processes.

Human Activities that Increase Soil Erosion

Agricultural Practices

Tilled farmland such as this is very susceptible to erosion from rainfall, due to the destruction of vegetative cover and the loosening of the soil during plowing.

Unsustainable agricultural practices are the single greatest contributor to the global increase in erosion rates. The tillage of agricultural lands, which breaks up soil into finer particles, is one of the primary factors. The problem has been exacerbated in modern times, due to mechanized agricultural equipment that allows for deep plowing, which severely increases the amount of soil that is available for transport by water erosion. Others include mono-cropping, farming on steep slopes, pesticide and chemical fertilizer usage (which kill organisms that bind soil together), row-cropping, and the use of surface irrigation. A complex overall situation with respect to defining nutrient losses from soils, could arise as a result of the size selective nature of soil erosion events. Loss of total phosphorus, for instance, in the finer eroded fraction is greater relative to the whole soil. Extrapolating this evidence to predict subsequent behaviour within receiving aquatic systems, the reason is that this more easily transported material may support a lower solution P concentration compared to coarser sized fractions. Tillage also increases wind erosion rates, by dehydrating the soil and breaking it up into smaller particles that can be picked up by the wind. Exacerbating this is the fact that most of the trees are generally removed from agricultural fields, allowing winds to have long, open runs to travel over at higher speeds. Heavy grazing reduces vegetative cover and causes severe soil compaction, both of which increase erosion rates.

Deforestation

In this clearcut, almost all of the vegetation has been stripped from surface of steep slopes, in an area with very heavy rains. Severe erosion occurs in cases such as this, causing stream sedimentation and the loss of nutrient rich topsoil.

In an undisturbed forest, the mineral soil is protected by a layer of *leaf litter* and an *humus* that cover the forest floor. These two layers form a protective mat over the soil that absorbs the impact of rain drops. They are porous and highly permeable to rainfall, and allow rainwater to slow percolate into the soil below, instead of flowing over the surface as runoff. The roots of the trees and plants hold together soil particles, preventing them from being washed away. The vegetative cover acts to reduce the velocity of the raindrops that strike the foliage and stems before hitting the ground, reducing their kinetic energy. However it is the forest floor, more than the canopy, that prevents surface erosion. The terminal velocity of rain drops is reached in about 8 metres (26 feet). Because forest canopies are usually higher than this, rain drops can often regain terminal velocity even after striking the canopy. However, the intact forest floor, with its layers of leaf litter and organic matter, is still able to absorb the impact of the rainfall.

Deforestation causes increased erosion rates due to exposure of mineral soil by removing the humus and litter layers from the soil surface, removing the vegetative cover that binds soil together, and causing heavy soil compaction from logging equipment. Once trees have been removed by fire or logging, infiltration rates become high and erosion low to the degree the forest floor remains intact. Severe fires can lead to significant further erosion if followed by heavy rainfall.

Globally one of the largest contributors to erosive soil loss in the year 2006 is the slash and burn treatment of tropical forests. In a number of regions of the earth, entire sectors of a country have been rendered unproductive. For example, on the Madagascar high central plateau, comprising approximately ten percent of that country's land area, virtually the entire landscape is sterile of vegetation, with gully erosive furrows typically in excess of 50 metres (160 ft) deep and 1 kilometre (0.6 miles) wide. Shifting cultivation is a farming system which sometimes incorporates the slash and burn method in some regions of the world. This degrades the soil and causes the soil to become less and less fertile.

Roads and Urbanization

Urbanization has major effects on erosion processes—first by denuding the land of vegetative cover, altering drainage patterns, and compacting the soil during construction; and next by covering the land in an impermeable layer of asphalt or concrete that increases the amount of surface runoff and increases surface wind speeds. Much of the sediment carried in runoff from urban areas (especially roads) is highly contaminated with fuel, oil, and other chemicals. This increased runoff, in addition to eroding and degrading the land that it flows over, also causes major disruption to surrounding watersheds by altering the volume and rate of water that flows through them, and filling them with chemically polluted sedimentation. The increased flow of water through local waterways also causes a large increase in the rate of bank erosion.

Climate Change

The warmer atmospheric temperatures observed over the past decades are expected to lead to a more vigorous hydrological cycle, including more extreme rainfall events. The rise in sea levels that has occurred as a result of climate change has also greatly increased coastal erosion rates.

Studies on soil erosion suggest that increased rainfall amounts and intensities will lead to greater rates of soil erosion. Thus, if rainfall amounts and intensities increase in many parts of the world as expected, erosion will also increase, unless amelioration measures are taken. Soil erosion rates are expected to change in response to changes in climate for a variety of reasons. The most direct is the change in the erosive power of rainfall. Other reasons include: a) changes in plant canopy caused by shifts in plant biomass production associated with moisture regime; b) changes in litter cover on the ground caused by changes in both plant residue decomposition rates driven by temperature and moisture dependent soil microbial activity as well as plant biomass production rates; c) changes in soil moisture due to shifting precipitation regimes and evapo-transpiration rates, which changes infiltration and runoff ratios; d) soil erodibility changes due to decrease in soil organic matter concentrations in soils that lead to a soil structure that is more susceptible to erosion and increased runoff due to increased soil surface sealing and crusting; e) a shift of winter precipitation from non-erosive snow to erosive rainfall due to increasing winter temperatures; f) melting of permafrost, which induces an erodible soil state from a previously non-erodible one; and g) shifts in land use made necessary to accommodate new climatic regimes.

Studies by Pruski and Nearing indicated that, other factors such as land use unconsidered, it is reasonable to expect approximately a 1.7% change in soil erosion for each 1% change in total precipitation under climate change.

Global Environmental Effects

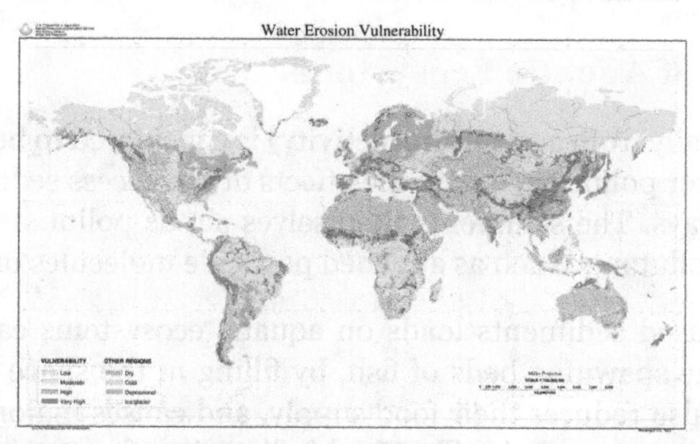

World map indicating areas that are vulnerable to high rates of water erosion.

During the 17th and 18th centuries, Easter Island experienced severe erosion due to deforestation and unsustainable agricultural practices. The resulting loss of topsoil ultimately led to ecological collapse, causing mass starvation and the complete disintegration of the Easter Island civilization.

Due to the severity of its ecological effects, and the scale on which it is occurring, erosion constitutes one of the most significant global environmental problems we face today.

Land Degradation

Water and wind erosion are now the two primary causes of land degradation; combined, they are responsible for 84% of degraded acreage.

Each year, about 75 billion tons of soil is eroded from the land—a rate that is about 13-40 times as fast as the natural rate of erosion. Approximately 40% of the world's agricultural land is seriously degraded. According to the United Nations, an area of fertile soil the size of Ukraine is lost every year because of drought, deforestation and climate change. In Africa, if current trends of soil degradation continue, the continent might be able to feed just 25% of its population by 2025, according to UNU's Ghana-based Institute for Natural Resources in Africa.

The loss of soil fertility due to erosion is further problematic because the response is often to apply chemical fertilizers, which leads to further water and soil pollution, rather than to allow the land to regenerate.

Sedimentation of Aquatic Ecosystems

Soil erosion (especially from agricultural activity) is considered to be the leading global cause of diffuse water pollution, due to the effects of the excess sediments flowing into the world's waterways. The sediments themselves act as pollutants, as well as being carriers for other pollutants, such as attached pesticide molecules or heavy metals.

The effect of increased sediments loads on aquatic ecosystems can be catastrophic. Silt can smother the spawning beds of fish, by filling in the space between gravel on the stream bed. It also reduces their food supply, and causes major respiratory issues for them as sediment enters their gills. The biodiversity of aquatic plant and algal life is reduced, and invertebrates are also unable to survive and reproduce. While the sedimentation event itself might be relatively short-lived, the ecological disruption caused by the mass die off often persists long into the future.

One of the most serious and long-running water erosion problems worldwide is in the People's Republic of China, on the middle reaches of the Yellow River and the upper reaches of the Yangtze River. From the Yellow River, over 1.6 billion tons of sediment flows into the ocean each year. The sediment originates primarily from water erosion in the Loess Plateau region of the northwest.

Airborne Dust Pollution

Soil particles picked up during wind erosion of soil are a major source of air pollution, in the form of airborne particulates—"dust". These airborne soil particles are often contaminated with toxic chemicals such as pesticides or petroleum fuels, posing ecological and public health hazards when they later land, or are inhaled/ingested.

Dust from erosion acts to suppress rainfall and changes the sky color from blue to white, which leads to an increase in red sunsets. Dust events have been linked to a decline in the health of coral reefs across the Caribbean and Florida, primarily since the 1970s. Similar dust plumes originate in the Gobi desert, which combined with pollutants, spread large distances downwind, or eastward, into North America.

Monitoring, Measuring and Modeling Soil Erosion

Terracing is an ancient technique that can significantly slow the rate of water erosion on cultivated slopes.

Monitoring and modeling of erosion processes can help people better understand the causes of soil erosion, make predictions of erosion under a range of possible conditions, and plan the implementation of preventative and restorative strategies for erosion. However, the complexity of erosion processes and the number of scientific disciplines that must be considered to understand and model them (e.g. climatology, hydrology, geology, soil science, agriculture, chemistry, physics, etc.) makes accurate modelling challenging. Erosion models are also non-linear, which makes them difficult to work with numerically, and makes it difficult or impossible to scale up to making predictions about large areas from data collected by sampling smaller plots.

The most commonly used model for predicting soil loss from water erosion is the *Universal Soil Loss Equation (USLE)*. This was developed in the 1960s and 1970s. It estimates the average annual soil loss A on a plot-sized area as:

$$A = RKLSCP$$

where R is the rainfall erosivity factor, K is the soil erodibility factor, L and S are topographic factors representing length and slope, C is the cover and management factor and P is the support practices factor.

Despite the USLE's plot-scale spatial focus, the model has often been used to estimate soil erosion on much larger areas, such as watersheds or even whole continents. For example, RUSLE has recently been used to quantify soil erosion across the whole of Europe . This is scientifically controversial, for several reasons. One major problem is that the USLE cannot simulate gully erosion, and so erosion from gullies is ignored in any USLE-based assessment of erosion. Yet erosion from gullies can be a substantial proportion (10-80%) of total erosion on cultivated and grazed land.

During the 50 years since the introduction of the USLE, many other soil erosion models have been developed. But because of the complexity of soil erosion and its constituent processes, all erosion models can give unsatisfactory results when validated i.e. when model predictions are compared with real-world measurements of erosion. Thus new soil erosion models continue to be developed. Some of these remain USLE-based, e.g. the G2 model . Other soil erosion models have largely (e.g. the Water Erosion Prediction Project model) or wholly (e.g. the Rangeland Hydrology and Erosion Model) abandoned usage of USLE elements.

Prevention and Remediation

A windbreak (the row of trees) planted next to an agricultural field, acting as a shield against strong winds. This reduces the effects of wind erosion, and provides many other benefits.

The most effective known method for erosion prevention is to increase vegetative cover on the land, which helps prevent both wind and water erosion. Terracing is an extremely effective means of erosion control, which has been practiced for thousands of years by people all over the world. Windbreaks (also called shelterbelts) are rows of trees and shrubs that are planted along the edges of agricultural fields, to shield the fields against winds. In addition to significantly reducing wind erosion, windbreaks provide many other benefits such as improved microclimates for crops (which are sheltered from the dehydrating and otherwise damaging effects of wind), habitat for beneficial bird species, carbon sequestration, and aesthetic improvements to the agricultural landscape. Traditional planting methods, such as mixed-cropping (instead of monocropping) and crop rotation have also been shown to significantly reduce erosion rates. Crop residues play a role in the mitigation of erosion, because they reduce the impact of raindrops breaking up the soil particles. There is a higher potential for erosion when producing potatoes than when growing cereals, or oilseed crops. Forages have a fibrous root system, which helps combat erosion by anchoring the plants to the top layer of the soil, and covering the entirety of the field, as it is a non-row crop.

Soil Retrogression and Degradation

Soil retrogression and degradation are two regressive evolution processes associated with the loss of equilibrium of a stable soil. Retrogression is primarily due to soil erosion and corresponds to a phenomenon where succession reverts the land to its natural physical state. Degradation is an evolution, different from natural evolution, related to the local climate and vegetation. It is due to the replacement of primary plant communities (known as climax vegetation) by the secondary communities. This replacement modifies the humus composition and amount, and affects the formation of the soil. It is directly related to human activity. Soil degradation may also be viewed as any change or ecological disturbance to the soil perceived to be deleterious or undesirable.

General

At the beginning of soil formation, the bare rock out crops is gradually colonized by pioneer species (lichens and mosses). They are succeeded by herbaceous vegetation, shrubs and finally forest. In parallel, the first humus-bearing horizon is formed (the A horizon), followed by some mineral horizons (B horizons). Each successive stage is characterized by a certain association of soil/vegetation and environment, which defines an ecosystem.

After a certain time of parallel evolution between the ground and the vegetation, a state of steady balance is reached. This stage of development is called climax by some ecologists and "natural potential" by others. Succession is the evolution towards climax. Regardless of its name, the equilibrium stage of primary succession is the highest natural form of development that the environmental factors are capable of producing.

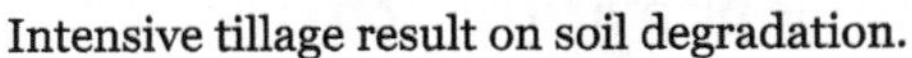

Intensive tillage result on soil degradation.

Willow hedge strengthened with fascines for the limitation of runoff, north of France.

The cycles of evolution of soils have very variable durations, between tens, hundreds, or thousands of years for quickly evolving soils (A horizon only) to more than a million years for slowly developing soils. The same soil may achieve several successive steady state conditions during its existence, as exhibited by the Pygmy forest sequence in Mendocino County, California. Soils naturally reach a state of high productivity, from which they naturally degrade as mineral nutrients are removed from the soil system. Thus older soils are more vulnerable to the effects of induced retrogression and degradation.

Ecological Factors Influencing Soil Formation

There are two types of ecological factors influencing the evolution of a soil (through alteration and humification). These two factors are extremely significant to explain the evolution of soils of short development.

- A first type of factor is the average climate of an area and the vegetation which is associated (biome).
- A second type of factor is more local, and is related to the original rock and local drainage. This type of factor explains appearance of specialized associations (ex peat bogs).

Biorhexistasy Theory

The destruction of the vegetation implies the destruction of evoluted soils, or a regressive evolution. Cycles of succession-regression of soils follow one another within short intervals of time (human actions) or long intervals of time (climate variations).

The climate role in the deterioration of the rocks and the formation of soils lead to the formulation of the theory of the biorhexistasy.

- In wet climate, the conditions are favorable to the deterioration of the rocks (mostly chemically), the development of the vegetation and the formation of soils; this period favorable to life is called biostasy.
- In dry climate, the rocks exposed are mostly subjected to mechanical disintegration which produces coarse detrital materials: this is referred to as rhexistasy.

Perturbations of the Balance of a Soil

When the state of balance, characterized by the ecosystem climax is reached, it tends to be maintained stable in the course of time. The vegetation installed on the ground provides the humus and ensures the ascending circulation of the matters. It protects the ground from erosion by playing the role of barrier (for example, protection from water and wind). Plants can also reduce erosion by binding the particles of the ground to their roots.

A disturbance of climax will cause retrogression, but often, secondary succession will start to guide the evolution of the system after that disturbance. Secondary succession is much faster than primary because the soil is already formed, although deteriorated and needing restoration as well.

However, when a significant destruction of the vegetation takes place (of natural origin such as an avalanche or human origin), the disturbance undergone by the ecosystem is too important. In this latter case, erosion is responsible for the destruction of the upper horizons of the ground, and is at the origin of a phenomenon of reversion to pioneer conditions. The phenomenon is called retrogression and can be partial or total (in this case, nothing remains beside bare rock). For example, the clearing of an inclined ground, subjected to violent rains, can lead to the complete destruction of the soil. Man can deeply modify the evolution of the soils by direct and brutal action, such as clearing, abusive cuts, forest pasture, litters raking. The climax vegetation is gradually replaced and the soil modified (example: replacement of leafy tree forests by moors or pines plantations). Retrogression is often related to very old human practices.

Influence of Human Activity

Soil erosion is the main factor for soil degradation and is due to several mechanisms: water erosion, wind erosion, chemical degradation and physical degradation.

Erosion is strongly related to human activity. For example, roads which increase impermeable surfaces lead to streaming and ground loss. Agriculture also accelerates soil erosion (increase of field size, correlated to hedges and ditches removal). Meadows are in regression to the profit of plowed lands. Spring cultures (sunflower, corn, beet) surfaces are increasing and leave the ground naked in winter. Sloping grounds are gradually colonized by vine. Lastly, use of herbicides leaves the ground naked between each crop. New cultural practices, such as mechanization also increases the risks of erosion. Fertilization by mineral manures rather than organic manure gradually destructure the soil. Many scientistsobserved a gradual decrease of soil organic matter content in soils, as well as a decrease of soil biological activity (in particular, in relation to chemical uses). Lastly, deforestation, in particular, is responsible for degradation of forest soils.

Agriculture increases the risk of erosion through its disturbance of vegetation by way of:

- Overgrazing of animals
- Monoculture planting
- Row cropping
- Tilling or plowing
- Crop removal
- Land-use conversion

Consequences of Soil Regression and Degradation

- yields impact: Recent increases in the human population have placed a great strain on the world's soil systems. More than 6 billion people are now using about 38% of the land area of the Earth to raise crops and livestock. Many soils suffer from various types of degradation, that can ultimately reduce their ability to produce food resources. Slight degradation refers to land where yield potential has been reduced by 10%, moderate degradation refers to a yield decrease from 10-50%. Severely degraded soils have lost more than 50% of their potential. Most severely degraded soils are located in developing countries.

- natural disasters: natural disasters such as mud flows, floods are responsible for the death of many living beings each year.

- deterioration of the water quality: the increase in the turbidity of water and the contribution of nitrogen and of phosphorus can result in eutrophication. Soils particles in surface waters are also accompanied by agricultural inputs and by some pollutants of industrial, urban and road origin (such as heavy metals). The ecological impact of agricultural inputs (such as weed killer) is known but difficult to evaluate because of the multiplicity of the products and their broad spectrum of action.

- biological diversity: soil degradation may involve perturbation of microbial communities, disappearance of the climax vegetation and decrease in animal habitat, thus leading to a biodiversity loss and animal extinction.

- economic loss: the estimated costs for land degradation are US $40 billion per year. This figure does not take into account negative externalities such as the increased use of fertilizers, loss of biodiversity and loss of unique landscapes.

Soil enhancement, Rebuilding, and Regeneration

Problems of soil erosion can be fought, and certain practices can lead to soil enhancement and rebuilding. Even though simple, methods for reducing erosion are often not chosen because these practices outweigh the short-term benefits. Rebuilding is especially possible through the improvement of soil structure, addition of organic matter and limitation of runoff. However, these techniques will never totally succeed to restore a soil (and the fauna and flora associated to it) that took more than 1000 years to build up. Soil regeneration is the reformation of degraded soil through biological, chemical, and or physical processes.

When productivity declined in the low-clay soils of northern Thailand, farmers initially responded by adding organic matter from termite mounds, but this was unsustainable in the long-term. Scientists experimented with adding bentonite, one of the smectite family of clays, to the soil. In field trials, conducted by scientists from the International Water Management Institute in cooperation with Khon Kaen University and local farmers, this had the effect of helping retain water and nutrients. Supplementing the farmer's usual practice with a single application of 200 kg bentonite per rai (6.26 rai = 1 hectare) resulted in an average yield increase of 73%. More work showed that applying bentonite to degraded sandy soils reduced the risk of crop failure during drought years.

In 2008, three years after the initial trials, IWMI scientists conducted a survey among 250 farmers in northeast Thailand, half who had applying bentonite to their fields and half who had not. The average output for those using the clay addition was 18% higher than for non-clay users. Using the clay had enabled some farmers to switch to growing vegetables, which need more fertile soil. This helped to increase their income. The researchers estimated that 200 farmers in northeast Thailand and 400 in Cambodia had adopted the use of clays, and that a further 20,000 farmers were introduced to the new technique.

References

- George, Rebecca; Joy, Varsha; S, Aiswarya; Jacob, Priya A. "Treatment Methods for Contaminated Soils - Translating Science into Practice" (PDF). International Journal of Education and Applied Research. Retrieved February 19, 2016.

4

Environmental Monitoring: Methods and Techniques

Environmental monitoring is the process that takes place in order to monitor the quality of the environment. The aspects elucidated in the section are air quality index, freshwater environmental quality parameters, water quality, oxygen saturation etc. The aspects elucidated in the section are of vital importance, and provide a better understanding of environmental monitoring.

Environmental Monitoring

Environmental monitoring describes the processes and activities that need to take place to characterise and monitor the quality of the environment. Environmental monitoring is used in the preparation of environmental impact assessments, as well as in many circumstances in which human activities carry a risk of harmful effects on the natural environment. All monitoring strategies and programmes have reasons and justifications which are often designed to establish the current status of an environment or to establish trends in environmental parameters. In all cases the results of monitoring will be reviewed, analysed statistically and published. The design of a monitoring programme must therefore have regard to the final use of the data before monitoring starts.

Air Quality Monitoring

Air quality monitoring is performed using specialized equipment and analytical methods used to establish air pollutant concentrations.

Air monitors are operated by citizens, regulatory agencies, and researchers to investigate air quality and the effects of air pollution.

Interpretation of ambient air monitoring data often involves a consideration of the spatial and temporal representativeness of the data gathered, and the health effects associated with exposure to the monitored levels.

Since air pollution is carried by the wind, consideration of anemometer data in the area between sources and the monitor often provides insights on the source of the air contaminants recorded by an air pollution monitor.

Close to the earth's surface, the atmosphere normally gets colder with height, but on certain days, the atmosphere begins to get warmer with height a short distance from the earth's surface, and air emissions build up under this "cap" on the vertical mixing.

Topographic features (such as a valley) that prevent lateral atmospheric mixing, coupled with the vertical cap on atmospheric mixing caused by an inversion, can lead to especially high air pollutant concentrations, for example, the 1948 Donora smog.

An inversion in a mountain valley in Poland

Air dispersion models that combine topographic, emissions and meteorological data to predict air pollutant concentrations are often helpful in interpreting air monitoring data.

If an air monitor produces concentrations of multiple chemical compounds, a unique "chemical fingerprint" of a particular air pollution source may emerge from analysis of the data.

Soil Monitoring

Soil monitoring is the process of collection of soil and testing in laboratory by analytical methods.

Soil sampling are of two types:

- Grab Sampling: in this method, sample is collected randomly from field
- Composite Sampling: In this method, mixing of multiple sub samples for larger and non-uniform fields.

In laboratory, soil can be tested for pH, Chlorides, Sulphates, Phosphates and other metals.

Water Quality Monitoring

Design of Environmental Monitoring Programmes

Water quality monitoring is of little use without a clear and unambiguous definition of the reasons for the monitoring and the objectives that it will satisfy. Almost all monitoring (except perhaps remote sensing) is in some part invasive of the environment under study and extensive and poorly planned monitoring carries a risk of damage to the environment. This may be a critical consideration in wilderness areas or when monitoring very rare organisms or those that are averse to human presence. Some monitoring techniques, such gill netting fish to estimate populations, can be very damaging, at least to the local population and can also degrade public trust in scientists carrying out the monitoring.

Almost all mainstream environmentalism monitoring projects form part of an overall monitoring strategy or research field, and these field and strategies are themselves derived from the high levels objectives or aspirations of an organisation. Unless individual monitoring projects fit into a wider strategic framework, the results are unlikely to be published and the environmental understanding produced by the monitoring will be lost.

Chemical

Analyzing water samples for pesticides

The range of chemical parameters that have the potential to affect any ecosystem is very large and in all monitoring programmes it is necessary to target a suite of parameters based on local knowledge and past practice for an initial review. The list can be expanded or reduced based on developing knowledge and the outcome of the initial surveys.

Freshwater environments have been extensively studied for many years and there is a robust understanding of the interactions between chemistry and the environment across much of the world. However, as new materials are developed and new pressures come to bear, revisions to monitoring programmes will be required. In the last 20 years acid rain, synthetic hormone analogues, halogenated hydrocarbons, greenhouse gases and many others have required changes to monitoring strategies.

Biological

In ecological monitoring, the monitoring strategy and effort is directed at the plants and animals in the environment under review and is specific to each individual study.

However, in more generalised environmental monitoring, many animals act as robust indicators of the quality of the environment that they are experiencing or have experienced in the recent past. One of the most familiar examples is the monitoring of numbers of Salmonid fish such as brown trout or Atlantic salmon in river systems and lakes to detect slow trends in adverse environmental effects. The steep decline in salmonid fish populations was one of the early indications of the problem that later became known as acid rain.

In recent years much more attention has been given to a more holistic approach in which the ecosystem health is assessed and used as the monitoring tool itself. It is this approach that underpins the monitoring protocols of the Water Framework Directive in the European Union.

Radiological

Radiation monitoring involves the measurement of radiation dose or radionuclide contamination for reasons related to the assessment or control of exposure to ionizing radiation or radioactive substances, and the interpretation of the results. The 'measurement' of dose often means the measurement of a dose equivalent quantity as a proxy (i.e. substitute) for a dose quantity that cannot be measured directly. Also, sampling may be involved as a preliminary step to measurement of the content of radionuclides in environmental media. The methodological and technical details of the design and operation of monitoring programmes and systems for different radionuclides, environmental media and types of facility are given in IAEA Safety Guide RS–G-1.8 and in IAEA Safety Report No. 64.

Radiation monitoring is often carried out using networks of fixed and deployable sensors such as the US Environmental Protection Agency's Radnet and the SPEEDI network in Japan. Airborne surveys are also made by organizations like the Nuclear Emergency Support Team.

Microbiological

Bacteria and viruses are the most commonly monitored groups of microbiological organisms and even these are only of great relevance where water in the aquatic environment is subsequently used as drinking water or where water contact recreation such as swimming or canoeing is practised.

Although pathogens are the primary focus of attention, the principal monitoring effort is almost always directed at much more common indicator species such as *Escherichia coli*, supplemented by overall coliform bacteria counts. The rationale behind this monitoring strategy is that most human pathogens originate from other humans via the sewage stream. Many sewage treatment plants have no sterilisation final stage and therefore discharge an effluent which, although having a clean appearance, still contains many millions of bacteria per litre, the majority of which are relatively harmless coliform bacteria. Counting the number of harmless (or less harmful) sewage bacteria allows a judgement to be made about the probability of significant numbers of pathogenic bacteria or viruses being present. Where *E. coli* or coliform levels exceed pre-set trigger values, more intensive monitoring including specific monitoring for pathogenic species is then initiated.

Populations

Monitoring strategies can produce misleading answers when relaying on counts of species or presence or absence of particular organisms if there is no regard to population size. Understanding the populations dynamics of an organism being monitored is critical.

As an example if presence or absence of a particular organism within a 10 km square is the measure adopted by a monitoring strategy, then a reduction of population from 10,000 per square to 10 per square will go unnoticed despite the very significant impact experienced by the organism.

Monitoring Programmes

All scientifically reliable environmental monitoring is performed in line with a published programme. The programme may include the overall objectives of the organisation, references to the specific strategies that helps deliver the objective and details of specific projects or tasks within those strategies. However the key feature of any programme is the listing of what is being monitored and how that monitoring is to take place and the time-scale over which it should all happen. Typically, and often as an appendix, a monitoring programme will provide a table of locations, dates and sampling methods that are proposed and which, if undertaken in full, will deliver the published monitoring programme.

There are a number of commercial software packages which can assist with the implementation of the programme, monitor its progress and flag up inconsistencies or omissions but none of these can provide the key building block which is the programme itself.

Environmental Monitoring Data Management Systems

Given the multiple types and increasing volumes and importance of monitoring data, commercial software Environmental Data Management Systems (EDMS) or E-MDMS are increasingly in common use by regulated industries. They provide a means of managing all monitoring data in a single central place. Quality validation, compliance checking, verifying all data has been received, and sending alerts are generally automated. Typical interrogation functionality enables comparison of data sets both temporarily and spatially. They will also generate regulatory and other reports.

Formal Certification:

(May 2014) there is only one certification scheme specifically for environmental data management software. This is provided by the Environment Agency in the UK under its Monitoring Certification Scheme (MCERTS) [Environmental certification].

Sampling Methods

There are a wide range of sampling methods which depend on the type of environment, the material being sampled and the subsequent analysis of the sample.

At its simplest a sample can be filling a clean bottle with river water and submitting it for conventional chemical analysis. At the more complex end, sample data may be produced by complex electronic sensing devices taking sub-samples over fixed or variable time periods.

Grab Samples

Collecting a grab sample on a stream

Grab samples are samples taken of a homogeneous material, usually water, in a single vessel. Filling a clean bottle with river water is a very common example. Grab samples provide a good snap-shot view of the quality of the sampled environment at the point of sampling and at the time of sampling. Without additional monitoring, the results cannot be extrapolated to other times or to other parts of the river, lake or ground-water.

In order to enable grab samples or rivers to be treated as representative, repeat transverse and longitudinal transect surveys taken at different times of day and times of year are required to establish that the grab-sample location is as representative as is reasonably possible. For large rivers such surveys should also have regard to the depth of the sample and how to best manage the sampling locations at times of flood and drought.

In lakes grab samples are relatively simple to take using depth samplers which can be lowered to a pre-determined depth and then closed trapping a fixed volume of water from the required depth. In all but the shallowest lakes, there are major changes in the chemical composition of lake water at different depths, especially during the summer months when many lakes stratify into a warm, well oxygenated upper layer (*epilimnion*) and a cool de-oxygenated lower layer *(hypolimnion*).

In the open seas marine environment grab samples can establish a wide range of base-line parameters such as salinity and a range of cation and anion concentrations. However, where changing conditions are an issue such as near river or sewage discharges, close to the effects of volcanism or close to areas of freshwater input from melting ice, a grab sample can only give a very partial answer when taken on its own.

Semi-continuous Monitoring and Continuous

An automated sampling station and data logger (to record temperature, specific conductance, and dissolved oxygen levels)

There is a wide range of specialized sampling equipment available that can be programmed to take samples at fixed or variable time intervals or in response to an external trigger. For example, a sampler can be programmed to start taking samples of a river at 8 minute intervals when the rainfall intensity rises above 1 mm / hour. The trigger in this case may be a remote rain gauge communicating with the sampler by using cell phone or meteor burst technology. Samplers can also take individual discrete samples at each sampling occasion or bulk up samples into composite so that in the course of one day, such a sampler might produce 12 composite samples each composed of 6 sub-samples taken at 20 minute intervals.

Continuous or quasi-continuous monitoring involves having an automated analytical facility close to the environment being monitored so that results can, if required, be viewed in real time. Such systems are often established to protect important water supplies such as in the River Dee regulation system but may also be part of an overall monitoring strategy on large strategic rivers where early warning of potential problems is essential. Such systems routinely provide data on parameters such as pH, dissolved oxygen, conductivity, turbidity and colour but it is also possible to operate gas liquid chromatography with mass spectrometry technologies (GLC/MS) to examine a wide range of potential organic pollutants. In all examples of automated bank-side analysis there is a requirement for water to be pumped from the river into the monitoring station. Choosing a location for the pump inlet is equally as critical as deciding on the location for a river grab sample. The design of the pump and pipework also requires careful design to avoid artefacts being introduced through the action of pumping the water. Dissolved oxygen concentration is difficult to sustain through a pumped system and GLC/MS facilities can detect micro-organic contaminants from the pipework and glands.

Passive Sampling

The use of passive samplers greatly reduces the cost and the need of infrastructure on the sampling location. Passive samplers are semi-disposable and can be produced at a relatively low cost, thus they can be employed in great numbers, allowing for a better cover and more data being collected. Due to being small the passive sampler can also be hidden, and thereby lower the risk of vandalism. Examples of passive sampling devices are the diffusive gradients in thin films (DGT) sampler, Chemcatcher, Polar organic chemical integrative sampler (POCIS), and an air sampling pump.

Remote Surveillance

Although on-site data collection using electronic measuring equipment is commonplace, many monitoring programmes also use remote surveillance and remote access to data in real time. This requires the on-site monitoring equipment to be connected to a base station via either a telemetry network,land-line, cell phone network or other telemetry system such as Meteor burst. The advantage of remote surveillance is that many data feeds can come into a single base station for storing and analysis. It also enable trigger levels or alert levels to be set for individual monitoring sites and/or parameters so that immediate action can be initiated if a trigger level is exceeded. The use of remote surveillance also allows for the installation of very discrete monitoring equipment which can often be buried, camouflaged or tethered at depth in a lake or river with only a short whip aerial protruding. Use of such equipment tends to reduce vandalism and theft when monitoring in locations easily accessible by the public.

Remote Sensing

Environmental remote sensing uses aircraft or satellites to monitor the environment using multi-channel sensors.

There are two kinds of remote sensing. Passive sensors detect natural radiation that is emitted or reflected by the object or surrounding area being observed. Reflected sunlight is the most common source of radiation measured by passive sensors and in environmental remote sensing, the sensors used are tuned to specific wavelengths from far infrared through visible light frequencies to the far ultraviolet. The volumes of data that can be collected are very large and require dedicated computational support . The output of data analysis from remote sensing are false colour images which differentiate small differences in the radiation characteristics of the environment being monitored. With a skilful operator choosing specific channels it is possible to amplify differences which are imperceptible to the human eye. In particular it is possible to discriminate subtle changes in chlorophyll a and chlorophyll b concentrations in plants and show areas of an environment with slightly different nutrient regimes.

Active remote sensing emits energy and uses a passive sensor to detect and measure the radiation that is reflected or backscattered from the target. LIDAR is often used to acquire information about the topography of an area, especially when the area is large and manual surveying would be prohibitively expensive or difficult.

Remote sensing makes it possible to collect data on dangerous or inaccessible areas. Remote sensing applications include monitoring deforestation in areas such as the Amazon Basin, the effects of climate change on glaciers and Arctic and Antarctic regions, and depth sounding of coastal and ocean depths.

Orbital platforms collect and transmit data from different parts of the electromagnetic spectrum, which in conjunction with larger scale aerial or ground-based sensing and analysis, provides information to monitor trends such as El Niño and other natural long and short term phenomena. Other uses include different areas of the earth sciences such as natural resource management, land use planning and conservation.

Bio-monitoring

The use of living organisms as monitoring tools has many advantages. Organisms living in the environment under study are constantly exposed to the physical, biological and chemical influences of that environment. Organisms that have a tendency to accumulate chemical species can often accumulate significant quantities of material from very low concentrations in the environment. Mosses have been used by many investigators to monitor heavy metal concentrations because of their tendency to selectively adsorb heavy metals.

Similarly, eels have been used to study halogenated organic chemicals, as these are adsorbed into the fatty deposits within the eel.

Other Sampling Methods

Ecological sampling requires careful planning to be representative and as noninvasive as possible. For grasslands and other low growing habitats the use of a quadrat – a 1-metre square frame – is often used with the numbers and types of organisms growing within each quadrat area counted

Sediments and soils require specialist sampling tools to ensure that the material recovered is representative. Such samplers are frequently designed to recover a specified volume of material and may also be designed to recover the sediment or soil living biota as well such as the Ekman grab sampler.

Data Interpretations

The interpretation of environmental data produced from a well designed monitoring programme is a large and complex topic addressed by many publications. Regrettably it is sometimes the case that scientists approach the analysis of results with a pre-conceived outcome in mind and use or misuse statistics to demonstrate that their own particular point of view is correct.

Statistics remains a tool that is equally easy to use or to misuse to demonstrate the lessons learnt from environmental monitoring.

Environmental Quality Indices

Since the start of science-based environmental monitoring, a number of quality indices have been devised to help classify and clarify the meaning of the considerable volumes of data involved. Stating that a river stretch is in "Class B" is likely to be much more informative than stating that this river stretch has a mean BOD of 4.2, a mean dissolved oxygen of 85%, etc. In the UK the Environment Agency formally employed a system called General Quality Assessment (GQA) which classified rivers into six quality letter bands from A to F based on chemical criteria and on biological criteria. The Environment Agency and its devolved partners in Wales (Countryside Council for Wales, CCW) and Scotland (Scottish Environmental Protection Agency, SEPA) now employ a system of biological, chemical and physical classification for rivers and lakes that corresponds with the EU Water Framework Directive.

Environmental Impact Assessment

Environmental assessment (EA) is the term used for the assessment of the environmental consequences (positive and negative) of a plan, policy, program, or concrete projects prior to the decision to move forward with the proposed action. In this context, the term "environmental impact assessment" (EIA) is usually used when applied to concrete projects by individuals or companies and the term "strategic environmental assessment" (SEA) applies to policies, plans and programmes most often proposed by organs of state (Fischer, 2016). Environmental assessments may be governed by rules of administrative procedure regarding public participation and documentation of decision making, and may be subject to judicial review.

The purpose of the assessment is to ensure that decision makers consider the environmental impacts when deciding whether or not to proceed with a project. The International Association for Impact Assessment (IAIA) defines an environmental impact assessment as "the process of identifying, predicting, evaluating and mitigating the biophysical, social, and other relevant effects of development proposals prior to major decisions being taken and commitments made." EIAs are unique in that they do not require adherence to a predetermined environmental outcome, but rather they require decision makers to account for environmental values in their decisions and to justify those decisions in light of detailed environmental studies and public comments on the potential environmental impacts.

CASI Global, New York. The Global Certification body for CSR & Sustainability goes one step further in this definition to include financial and supply chain effects. Any effects of the projects supply chain on the environment, ecology, human resources or country resources are to be included in the impact analysis. Most Global Fellows of CASI NY have to write a paper on the same and are authorized to undertake such projects in their country.

Engineering and consulting companies work hand in hand as contractors for mining, energy, oil & gas companies executing EIAs. Companies operating globally such as Arcadis, Royal HaskoningDHV, Golder Associates, Amec Foster Wheeler, Schlumberger Water Services (a Schlumberger company), ERM are an example of a much bigger pool of expertise globally. These contractors are the ones not only in charge of preparing an EIA study but most importantly getting these studies approved by each country government offices prior to the execution of a project. Each country will also have its own local contractors offering the same kind of service hence breaking out monopolies by increasing the supply of EIAs execution consultants.

History

Environmental impact assessments commenced in the 1960s, as part of increasing environmental awareness. EIAs involved a technical evaluation intended to contribute to more objective decision making. In the United States, environmental impact assessments obtained formal status in 1969, with enactment of the National Environmental Policy Act. EIAs have been used increasingly around the world. The number of "Environmental Assessments" filed every year "has vastly overtaken the number of more rigorous Environmental Impact Statements (EIS)." An Environmental Assessment is a "mini-EIS designed to provide sufficient information to allow the agency to decide whether the preparation of a full-blown Environmental Impact Statement (EIS) is necessary." EIA is an activity that is done to find out the impact that would be done before development will occur.

Methods

General and industry specific assessment methods are available including:

- *Industrial products* - Product environmental life cycle analysis (LCA) is used for

- identifying and measuring the impact of industrial products on the environment. These EIAs consider activities related to extraction of raw materials, ancillary materials, equipment; production, use, disposal and ancillary equipment.
- *Genetically modified plants* - Specific methods available to perform EIAs of genetically modified organisms include GMP-RAM and INOVA.
- *Fuzzy logic* - EIA methods need measurement data to estimate values of impact indicators. However, many of the environment impacts cannot be quantified, e.g. landscape quality, lifestyle quality and social acceptance. Instead information from similar EIAs, expert judgment and community sentiment are employed. Approximate reasoning methods known as fuzzy logic can be used. A fuzzy arithmetic approach has also been proposed and implemented using a software tool (TDEIA).

Follow-up

At the end of the project, an audit evaluates the accuracy of the EIA by comparing actual to predicted impacts. The objective is to make future EIAs more valid and effective. Two primary considerations are:

- *Scientific* - to examine the accuracy of predictions and explain errors
- *Management* - to assess the success of mitigation in reducing impacts

Audits can be performed either as a rigorous assessment of the null hypothesis or with a simpler approach comparing what actually occurred against the predictions in the EIA document.

After an EIA, the precautionary and polluter pays principles may be applied to decide whether to reject, modify or require strict liability or insurance coverage to a project, based on predicted harms.

The Hydropower Sustainability Assessment Protocol is a sector specific method for checking the quality of Environmental and Social assessments and management plans.

Around the World

Australia

The history of EIA in Australia could be linked to the enactment of the U.S. National Environment Policy Act (NEPA) in 1970, which made the preparation of environmental impact statements a requirement. In Australia, one might say that the EIA procedures were introduced at a State Level prior to that of the Commonwealth (Federal), with a majority of the states having divergent views to the Commonwealth. One of the pioneering states was New South Wales, whose State Pollution Control Commission issued EIA guidelines in 1974. At a Commonwealth (Federal) level, this was followed by passing of the Environment Protection (Impact of Proposals) Act in 1974. The Environment Protection and Biodiversity Conservation Act 1999 (EPBC) superseded the Environment Protection (Impact of Proposals) Act 1974 and is the current central piece for EIA in Australia on a Commonwealth (Federal) level. An important point to note is that this Commonwealth Act does not affect the validity of the States and Territories environmental and development assessments and approvals; rather the EPBC runs as a parallel to the State/Territory Systems. Overlap between federal and state requirements is addressed via bilateral agreements or one off accreditation of state processes, as provided for in the EPBC Act.

The Commonwealth Level

The EPBC Act provides a legal framework to protect and manage nationally and internationally important flora, fauna, ecological communities and heritage places- defined in the EPBC Act as matters of 'national environmental significance'. Following are the eight matters of 'national environmental significance' to which the EPBC ACT applies:

- World Heritage sites
- National Heritage places
- RAMSAR wetlands of international significance
- Listed threatened species and ecological communities
- Migratory species protected under international agreements
- The Commonwealth marine environment
- Nuclear actions (including uranium mining)
- National Heritage.

In addition to this, the EPBC Act aims at providing a streamlined national assessment and approval process for activities. These activities could be by the Commonwealth, or its agents, anywhere in the world or activities on Commonwealth land; and activities that are listed as having a 'significant impact' on matters of 'national environment significance'.

The EPBC Act comes into play when a person (a 'proponent') wants an action (often called a 'proposal' or 'project') assessed for environmental impacts under the EPBC Act, he or she must refer the project to the Department of Environment, Water, Heritage and the Arts (Australia). This 'referral' is then released to the public, as well as relevant state, territory and Commonwealth ministers, for comment on whether the project is likely to have a significant impact on matters of national environmental significance. The Department of Environment, Water, Heritage and the Arts assess the process and makes recommendation to the minister or the delegate for the feasibility. The final discretion on the decision remains of the minister, which is not solely based on matters of 'national environmental significance' but also the consideration of social and economic impact of the project.

The Australian Government environment minister cannot intervene in a proposal if it has no significant impact on one of the eight matters of 'national environmental significance' despite the fact that there may be other undesirable environmental impacts. This is primarily due to the division of powers between the States and the Federal government and due to which the Australian Government environment minister cannot overturn a state decision.

There are strict civil and criminal penalties for the breach of EPBC Act. Depending on the kind of breach, civil penalty (maximum) may go up to $550,000 for an individual and $5.5 million for a body corporate, or for criminal penalty (maximum) of seven years imprisonment and/or penalty of $46,200.

The State and Territory Level

Australian Capital Territory (ACT)

EIA provisions within Ministerial Authorities in the ACT are found in the Chapters 7 and 8 of the *Planning and Development Act 2007* (ACT). EIA in ACT was previously administered with the help of Part 4 of the Land (Planning and Environment) Act 1991 (Land Act) and Territory Plan (plan for land-use). Note that some EIA may occur in the ACT on Commonwealth land under the EPBC Act (Cth). Further provisions of the *Australian Capital Territory (Planning and Land Management) Act 1988* (Cth) may also be applicable particularly to national land and "designated areas".

New South Wales (NSW)

In New South Wales, the Environment Planning Assessment Act 1979 (EPA) establishes three pathways for EIA. The first is under Part 5.1 of the EPAA, which provides for EIA of 'State Significant Infrastructure' projects. (From June 2011, this Part replaced Part 3A, which previously covered EIA of major projects). The second is under Part 4 of the Act dealing with development control. If a project does not require approval under Part 3A or Part 4 it is then potentially captured by the third pathway, Part 5 dealing with environment impact assessment.

Northern Territory (NT)

The EIA process in Northern Territory is chiefly administered under the Environmental Assessment Act (EAA). Although EAA is the primary tool for EIA in Northern Territory, there are further provisions for proposals in the Inquiries Act 1985 (NT).

Queensland (QLD)

There are four main EIA processes in Queensland. Firstly, under the Integrated Planning Act 1997 (IPA) for development projects other than mining. Secondly, under the Environmental Protection Act 1994 (EP Act) for some mining and petroleum activities. Thirdly, under the State Development and Public Works Organization Act 1971 (State Development Act) for 'significant projects'. Finally, Environment Protection and Biodiversity Conservation Act 1999 (Cth) for 'controlled actions'.

South Australia (SA)

The local governing tool for EIA in South Australia is the Development Act 1993. There are three levels of assessment possible under the Act in the form of an environment impact statement (EIS), a public environmental report (PER) or a Development Report (DR).

Tasmania (TAS)

In Tasmania, an integrated system of legislation is used to govern development and approval process, this system is a mixture of the Environmental Management and Pollution Control Act 1994 (EMPCA), Land Use Planning and Approvals Act 1993 (LUPAA), State Policies and Projects Act 1993 (SPPA), and Resource Management and Planning Appeals Tribunal Act 1993.

Victoria (VIC)

The EIA process in Victoria is intertwined with the Environment Effects Act 1978 and the Ministerial Guidelines for Assessment of Environmental Effects (made under the s. 10 of the EE Act).

Western Australia (WA)

The Environmental Protection Act 1986 (Part 4) provides the legislative framework for the EIA process in Western Australia. The EPA Act oversees the planning and development proposals and assesses their likely impacts on the environment.

Canada

In *Friends of the Oldman River Society v. Canada (Minister of Transportation)*,(SCC 1992) La Forest J of the Supreme Court of Canada described environmental impact assessment in terms of the proper scope of federal jurisdiction with respect to environments matters,

"Environmental impact assessment is, in its simplest form, a planning tool that is now generally regarded as an integral component of sound decision-making."

Supreme Court Justice La Forest cited (Cotton, Emond & 1981 245), "The basic concepts behind environmental assessment are simply stated: (1) early identification and evaluation of all potential environmental consequences of a proposed undertaking; (2) decision making that both guarantees the adequacy of this process and reconciles, to the greatest extent possible, the proponent's development desires with environmental protection and preservation."

La Forest referred to (Jeffrey 1989, 1.2,1.4) and (Emond 1978, p. 5) who described "... environmental assessments as a planning tool with both an information-gathering and a decision-making component" that provide "...an objective basis for granting or denying approval for a proposed development."

Justice La Forest addressed his concerns about the implications of Bill C-45 regarding public navigation rights on lakes and rivers that would contradict previous cases.(La Forest & 1973 178-80)

The Canadian Environmental Assessment Act 2012 (CEAA 2012) "and its regulations establish the legislative basis for the federal practice of environmental assessment in most regions of Canada." CEAA 2012 came into force July 6, 2012 and replaces the former *Canadian Environmental Assessment Act* (1995). EA is defined as a planning tool to identify, understand, assess and mitigate, where possible, the environmental effects of a project.

"The purposes of this Act are: (a) to protect the components of the environment that are within the legislative authority of Parliament from significant adverse environmental effects caused by a designated project; (b) to ensure that designated projects that require the exercise of a power or performance of a duty or function by a federal authority under any Act of Parliament other than this Act to be carried out, are considered in a careful and precautionary manner to avoid significant adverse environmental effects; (c) to promote cooperation and coordinated action between federal and provincial governments with respect to environmental assessments; (d) to promote communication and cooperation with aboriginal peoples with respect to environmental assessments; (e) to ensure that opportunities are provided for meaningful public participation during an environmental assessment; (f) to ensure that an environmental assessment is completed in a timely manner; (g) to ensure that projects, as defined in section 66, that are to be carried out on federal lands, or those that are outside Canada and that are to be carried out or financially supported by a federal authority, are considered in a careful and precautionary manner to avoid significant adverse environmental effects; (h) to encourage federal authorities to take actions that promote sustainable development in order to achieve or maintain a healthy environment and a healthy economy; and (i) to encourage the study of the cumulative effects of physical activities in a region and the consideration of those study results in environmental assessments."

Opposition

Environmental Lawyer Dianne Saxe argued that the CEAA 2012 "allows the federal government to create mandatory timelines for assessments of even the largest and most important projects, regardless of public opposition." (Saxe 2012)

"Now that federal environmental assessments are gone, the federal government will only assess very large, very important projects. But it's going to do them in a hurry."

Dianne Saxe

On 3 August 2012 the Canadian Environmental Assessment Agency nine "designated projects" with their timelines: Enbridge Northern Gateway Pipeline Joint Review Panel (JRP) 18 months; Marathon Platinum Group Metals and Copper Mine Project (JRP): 13 months; Site C Clean Energy Project (JRP) 8.5 months; Deep Geologic Repository Project (JRP) 17 months; Enbridge Northern Gateway Project (JRP) 18 months; Jackpine Mine Expansion Project (JRP) 11.5 months; Pierre River Mine Project: 8 months; New Prosperity Gold-Copper Mine Project (JRP) 7.5 months; Frontier Oil Sands Mine Project (JRP)8.5 months; EnCana/Cenovus Shallow Gas Infill Project (JRP) 5 months.

Saxe compares these timelines with environmental assessments for the Mackenzie Valley Pipeline. Thomas R. Berger, Royal Commissioner of the Mackenzie Valley Pipeline Inquiry (9 May 1977), worked extremely hard to ensure that industrial development on Aboriginal people's land resulted in benefits to those indigenous people.

On 22 April 2013, Official Opposition Environment critic Megan Leslie issued a statement claiming that the federal government's recent changes to "fish habitat protection, the Navigable Waters Protection Act and the Canadian Environmental Assessment Act", along with gutting existing laws and making cuts to science and research, "will be disastrous, not only for the environment, but also for Canadians' health and economic prosperity." On 26 September 2012, Leslie argued that with the changes to the Canadian Environmental Assessment Act that came into effect 6 July 2012, "seismic testing, dams, wind farms and power plants" no longer required any federal environmental assessment. She also claimed that because the CEAA 2012—which she claimed was rushed through Parliament—dismantled the CEAA 1995, the Oshawa ethanol plant project would no longer have a full federal environmental assessment. Mr. Peter Kent (Minister of the Environment) explained that the CEAA 2012 "provides for the Government of Canada and the Environmental Assessment Agency to focus on the large and most significant projects that are being proposed across the country." The 2,000 to 3,000-plus smaller screenings that were in effect under CEAA 1995 became the "responsibility of lower levels of government but are still subject to the same strict federal environmental laws." Anne Minh-Thu Quach, MP for Beauharnois—Salaberry, QC, argued that the mammoth budget bill dismantled 50 years of environmental protection without consulting Canadians about the "colossal changes they are making to environmental assessments." She claimed that the federal government is entering into "limited consultations, by invitation only, months after the damage was done."

China

The Environmental Impact Assessment Law (EIA Law) requires that an environmental impact assessment be completed prior to project construction. However, if a developer completely ignores this requirement and builds a project without submitting an environmental impact statement, the only penalty is that the environmental protection bureau (EPB) may require the developer to do a make-up environmental assessment. If the developer does not complete this make-up assessment within the designated time, only then is the EPB authorized to fine the developer. Even so, the possible fine is capped at a maximum of about US$25,000, a fraction of the overall cost of most major projects. The lack of more stringent enforcement mechanisms has resulted in a significant percentage of projects not completing legally required environmental impact assessments prior to construction.

China's State Environmental Protection Administration (SEPA) used the legislation to halt 30 projects in 2004, including three hydro-power plants under the Three Gorges Project Company. Although one month later (Note as a point of reference, that the typical EIA for a major project in the USA takes one to two years.), most of the 30 halted projects resumed their construction, reportedly having passed the environmental assessment, the fact that these key projects' construction was ever suspended was notable.

A joint investigation by SEPA and the Ministry of Land and Resources in 2004 showed that 30-40% of the mining construction projects went through the procedure of environment impact assessment as required, while in some areas only 6-7% did so. This partly explains why China has witnessed so many mining accidents in recent years.

SEPA alone cannot guarantee the full enforcement of environmental laws and regulations, observed Professor Wang Canfa, director of the centre to help environmental victims at China University of Political Science and Law. In fact, according to Wang, the rate of China's environmental laws and regulations that are actually enforced is estimated at barely 10%.

Egypt

Environmental Impact Assessment (EIA) EIA is implemented in Egypt under the umbrella of the Ministry of state for environmental affairs. The Egyptian Environmental Affairs Agency (EEAA) is responsible for the EIA services.

In June 1997, the responsibility of Egypt's first full-time Minister of State for Environmental Affairs was assigned as stated in the Presidential Decree no.275/1997.

From thereon, the new ministry has focused, in close collaboration with the national and international development partners, on defining environmental policies, setting priorities and implementing initiatives within a context of sustainable development.

According to the Law 4/1994 for the Protection of the Environment, the Egyptian Environmental Affairs Agency (EEAA) was restructured with the new mandate to substitute the institution initially established in 1982. At the central level, EEAA represents the executive arm of the Ministry.

The purpose of EIA is to ensure the protection and conservation of the environment and natural resources including human health aspects against uncontrolled development. The long-term objective is to ensure a sustainable economic development that meets present needs without compromising future generations ability to meet their own needs. EIA is an important tool in the integrated environmental management approach.

EIA must be performed for new establishments or projects and for expansions or renovations of existing establishments according to the Law for the Environment.

EU

There is a wide range of instruments in the Environmental policy of the European Union. Among them the European Union has established a mix of mandatory and discretionary procedures to assess environmental impacts. European Union Directive (85/337/EEC) on Environmental Impact Assessments (known as the *EIA Directive*) was first introduced in 1985 and was amended in 1997. The directive was amended again in 2003, following EU signature of the 1998 Aarhus Convention, and once more in 2009. The initial Directive of 1985 and its three amendments have been codified in Directive 2011/92/EU of 13 December 2011. In 2001, the issue was enlarged to the assessment of plans and programmes by the so-called *Strategic Environmental Assessment (SEA) Directive* (2001/42/EC), which is now in force. Under the EU directive, an EIA must provide certain information to comply. There are seven key areas that are required:

1. Description of the project
    - Description of actual project and site description
    - Break the project down into its key components, i.e. construction, operations, decommissioning
    - For each component list all of the sources of environmental disturbance
    - For each component all the inputs and outputs must be listed, e.g., air pollution, noise, hydrology
2. Alternatives that have been considered
    - Examine alternatives that have been considered

- Example: in a biomass power station, will the fuel be sourced locally or nationally?

3. Description of the environment

   - List of all aspects of the environment that may be affected by the development
   - Example: populations, fauna, flora, air, soil, water, humans, landscape, cultural heritage
   - This section is best carried out with the help of local experts, e.g. the RSPB in the UK

4. Description of the significant effects on the environment

   - The word significant is crucial here as the definition can vary
   - 'Significant' must be defined
   - The most frequent method used here is use of the Leopold matrix
   - The matrix is a tool used in the systematic examination of potential interactions
   - Example: in a windfarm development a significant impact may be collisions with birds

5. Mitigation

   - This is where EIA is most useful
   - Once section 4 is complete, it is obvious where impacts are greatest
   - Using this information ways to avoid negative impacts should be developed
   - Best working with the developer with this section as they know the project best
   - Using the windfarm example again construction could be out of bird nesting seasons

6. Non-technical summary (EIS)

   - The EIA is in the public domain and be used in the decision making process
   - It is important that the information is available to the public

- This section is a summary that does not include jargon or complicated diagrams
- It should be understood by the informed lay-person

7. Lack of know-how/technical difficulties
    - This section is to advise any areas of weakness in knowledge
    - It can be used to focus areas of future research
    - Some developers see the EIA as a starting block for poor environmental management

Annexed Projects

All projects are either classified as Annex 1 or Annex 2 projects. Those lying in Annex 1 are large scale developments such as motorways, chemical works, bridges, powerstations etc. These always require an EIA under the Environmental Impact Assessment Directive (85,337,EEC as amended). Annex 2 projects are smaller in scale than those referred to in Annex 1. Member States must determine whether these project shall be made subject to an assessment subject to a set of criteria set out in Annex 3 of codified Directive 2011/92/EU.

The Netherlands

EIA was implemented in Dutch legislation on September 1, 1987. The categories of projects that require an EIA are summarised in Dutch legislation, the Wet milieubeheer. The use of thresholds for activities makes sure that EIA is obligatory for those activities that may have considerable impacts on the environment.

For projects and plans that fit these criteria, an EIA report is required. The EIA report defines a.o. the proposed initiative, it makes clear the impact of that initiative on the environment and compares this with the impact of possible alternatives with less a negative impact.

Hong Kong

EIA in Hong Kong, since 1998, is regulated by the *Environmental Impact Assessment Ordinance 1997.*

The original proposal to construct the Lok Ma Chau Spur Line overground across the Long Valley failed to get through EIA, and the Kowloon–Canton Railway Corporation had to change its plan and build the railway underground. In April 2011, the EIA of the Hong Kong section of the Hong Kong-Zhuhai-Macau Bridge was found to have breached the ordinance, and was declared unlawful. The appeal by the government was allowed in September 2011. However, it was estimated that this EIA court case had increased the construction cost of the Hong Kong section of the bridge by HK$6.5 billion in money-of-the-day prices.

India

The Ministry of Environment, Forests and Climate Change (MoEFCC) of India has been in a great effort in Environmental Impact Assessment in India. The main laws in action are the Water Act(1974), the Indian Wildlife (Protection) Act (1972), the Air (Prevention and Control of Pollution) Act (1981) and the Environment (Protection) Act (1986),Biological Diversity Act(2002). The responsible body for this is the Central Pollution Control Board. Environmental Impact Assessment (EIA) studies need a significant amount of primary and secondary environmental data. Primary data are those collected in the field to define the status of the environment (like air quality data, water quality data etc.). Secondary data are those collected over the years that can be used to understand the existing environmental scenario of the study area. The environmental impact assessment (EIA) studies are conducted over a short period of time and therefore the understanding of the environmental trends, based on a few months of primary data, has limitations. Ideally, the primary data must be considered along with the secondary data for complete understanding of the existing environmental status of the area. In many EIA studies, the secondary data needs could be as high as 80% of the total data requirement. EIC is the repository of one stop secondary data source for environmental impact assessment in India.

The Environmental Impact Assessment (EIA) experience in India indicates that the lack of timely availability of reliable and authentic environmental data has been a major bottle neck in achieving the full benefits of EIA. The environment being a multi-disciplinary subject, a multitude of agencies are involved in collection of environmental data. However, no single organization in India tracks available data from these agencies and makes it available in one place in a form required by environmental impact assessment practitioners. Further, environmental data is not available in enhanced forms that improve the quality of the EIA. This makes it harder and more time-consuming to generate environmental impact assessments and receive timely environmental clearances from regulators. With this background, the Environmental Information Centre (EIC) has been set up to serve as a professionally managed clearing house of environmental information that can be used by MoEF, project proponents, consultants, NGOs and other stakeholders involved in the process of environmental impact assessment in India. EIC caters to the need of creating and disseminating of organized environmental data for various developmental initiatives all over the country.

EIC stores data in GIS format and makes it available to all environmental impact assessment studies and to EIA stakeholders in a cost effective and timely manner. So that we can manage that in different proportions such as remedy measures etc.,

Korea, South

Recycling culture and policy Ministry of Environment

Malaysia

In Malaysia, Section 34A, Environmental Quality Act, 1974 requires developments that have significant impact to the environment are required to conduct the Environmental impact assessment.

Nepal

In Nepal, EIA has been integrated in major development projects since the early 1980s. In the planning history of Nepal, the sixth plan (1980–85), for the first time, recognized the need for EIA with the establishment of Environmental Impact Study Project (EISP) under the Department of Soil Conservation in 1982 to develop necessary instruments for integration of EIA in infrastructure development projects. However, the government of Nepal enunciated environment conservation related policies in the seventh plan (NPC, 1985–1990). To enforce this policy and make necessary arrangements, a series of guidelines were developed, thereby incorporating the elements of environmental factors right from the project formulation stage of the development plans and projects and to avoid or minimize adverse effects on the ecological system. In addition, it has also emphasized that EIAs of industry, tourism, water resources, transportation, urbanization, agriculture, forest and other developmental projects be conducted.

In Nepal, the government's Environmental Impact Assessment Guideline of 1993 inspired the enactment of the Environment Protection Act (EPA) of 1997 and the Environment Protection Rules (EPR) of 1997 (EPA and EPR have been enforced since 24 and 26 June 1997 respectively in Nepal) to internalizing the environmental assessment system. The process institutionalized the EIA process in development proposals and enactment, which makes the integration of IEE and EIA legally binding to the prescribed projects. The projects, requiring EIA or IEE, are included in Schedules 1 and 2 of the EPR, 1997 (GoN/MoLJPA 1997). Progresses were made in the Environmental protection issue during the 8th five-year plan (1992–1997). The following development in Environmental protection were achieved during that time:

- Formulation of Environmental Protection Act 1997
- Establishment of Ministry of Environment
- Development of National Environmental Policies and Action Plan, EIA guidelines developed
- Consideration of environmental concerns in hydropower projects
- Development of industrial, irrigation and agricultural policies that undertook environmental concerns

New Zealand

In New Zealand, EIA is usually referred to as *Assessment of Environmental Effects* (AEE). The first use of EIA's dates back to a Cabinet minute passed in 1974 called Environmental Protection and Enhancement Procedures. This had no legal force and only related to the activities of government departments. When the Resource Management Act was passed in 1991, an EIA was required as part of a resource consent application. Section 88 of the Act specifies that the AEE must include "such detail as corresponds with the scale and significance of the effects that the activity may have on the environment". While there is no duty to consult any person when making a resource consent application (Sections 36A and Schedule 4), proof of consultation is almost certain required by local councils when they decide whether or not to publicly notify the consent application under Section 93.

Russian Federation

As of 2004, the state authority responsible for conducting the State EIA in Russia has been split between two Federal bodies: 1) Federal service for monitoring the use of natural resources – a part of the Russian Ministry for Natural Resources and Environment and 2) Federal Service for Ecological, Technological and Nuclear Control. The two main pieces of environmental legislation in Russia are: The Federal Law 'On Ecological Expertise, 1995 and the 'Regulations on Assessment of Impact from Intended Business and Other Activity on Environment in the Russian Federation, 2000.

Federal Service for Monitoring the Use of Natural Resources

In 2006, the parliament committee on ecology in conjunction with the Ministry for Natural Resources and Environment, created a working group to prepare a number of amendments to existing legislation to cover such topics as stringent project documentation for building of potentially environmentally damaging objects as well as building of projects on the territory of protected areas. There has been some success in this area, as evidenced from abandonment of plans to construct a gas pipe-line through the only remaining habitat of the critically endangered Amur leopard in the Russian Far East.

Federal Service for Ecological, Technological and Nuclear Control

The government's decision to hand over control over several important procedures, including state EIA in the field of all types of energy projects, to the Federal Service for Ecological, Technological and Nuclear Control had caused a major controversy and criticism from environmental groups that blamed the government for giving nuclear power industry control over the state EIA.

Not surprisingly the main problem concerning State EIA in Russia is the clear differentiation of jurisdiction between the two above-mentioned Federal bodies.

Sri Lanka

The National Environmental Act, 1998 requires environmental impact assessment for large scale projects in sensitive areas. It is enforced by the Central Environmental Authority.

United States

The National Environmental Policy Act of 1969 (NEPA), enacted in 1970, established a policy of environmental impact assessment for federal agency actions, federally funded activities or federally permitted/licensed activities that in the U. S. is termed "environmental review" or simply "the NEPA process." The law also created the Council on Environmental Quality, which promulgated regulations to codify the law's requirements. Under United States environmental law an Environmental Assessment (EA) is compiled to determine the need for an *Environmental Impact Statement* (EIS). Federal or federalized actions expected to subject or be subject to significant environmental impacts will publish a Notice of Intent to Prepare an EIS as soon as significance is known. Certain actions of federal agencies must be preceded by the NEPA process. Contrary to a widespread misconception, NEPA does not prohibit the federal government or its licensees/permittees from harming the environment, nor does it specify any penalty if an environmental impact assessment turns out to be inaccurate, intentionally or otherwise. NEPA requires that plausible statements as to the prospective impacts be disclosed in advance. The purpose of NEPA process is to ensure that the decision maker is fully informed of the environmental aspects and consequences prior to making the final decision.

Environmental Assessment

An environmental assessment (EA) is an environmental analysis prepared pursuant to the National Environmental Policy Act to determine whether a federal action would significantly affect the environment and thus require a more detailed *Environmental Impact Statement (EIS)*. The certified release of an Environmental Assessment results in either a *Finding of No Significant Impact (FONSI)* or an EIS.

The Council on Environmental Quality (CEQ), which oversees the administration of NEPA, issued regulations for implementing the NEPA in 1979. Eccleston reports that the NEPA regulations barely mention preparation of EAs. This is because the EA was originally intended to be a simple document used in relatively rare instances where an agency was not sure if the potential significance of an action would be sufficient to trigger preparation of an EIS. But today, because EISs are so much longer and complicated to prepare, federal agencies are going to great effort to avoid preparing EISs by using EAs, even in cases where the use of EAs may be inappropriate. The ratio of EAs that are being issued compared to EISs is about 100 to 1.

Likewise, even the preparation of an accurate EA is viewed today as an onerous burden by many entities responsible for the environmental review of a proposal. Federal agencies have responded by streamlining their regulations that implement NEPA environmental review, by defining categories of projects that by their well understood nature may be safely excluded from review under NEPA, and by drawing up lists of project types that have negligible material impact upon the environment and can thus be exempted.

Content

The Environmental Assessment is a concise public document prepared by the federal action agency that serves to:

1. briefly provide sufficient evidence and analysis for determining whether to prepare an EIS or a Finding of No Significant Impact (FONSI)
2. Demonstrate compliance with the act when no EIS is required
3. facilitate the preparation of an EIS when a FONSI cannot be demonstrated

The Environmental Assessment includes a brief discussion of the purpose and need of the proposal and of its alternatives as required by NEPA 102(2)(E), and of the human environmental impacts resulting from and occurring to the proposed actions and alternatives considered practicable, plus a listing of studies conducted and agencies and stakeholders consulted to reach these conclusions. The action agency must approve an EA before it is made available to the public. The EA is made public through notices of availability by local, state, or regional clearing houses, often triggered by the purchase of a public notice advertisement in a newspaper of general circulation in the proposed activity area.

Structure

The structure of a generic Environmental Assessment is as follows:

1. Summary
2. Introduction
    - Background
    - Purpose and Need for Action
    - Proposed Action
    - Decision Framework
    - Public Involvement
    - Issues

3. Alternatives, including the Proposed Action
    - Alternatives
    - Mitigation Common to All Alternatives
    - Comparison of Alternatives
4. Environmental Consequences
5. Consultation and Coordination

Procedure

The EA becomes a draft public document when notice of it is published, usually in a newspaper of general circulation in the area affected by the proposal. There is a 15-day review period required for an Environmental Assessment (30 days if exceptional circumstances) while the document is made available for public commentary, and a similar time for any objection to improper process. Commenting on the Draft EA is typically done in writing or email, submitted to the lead action agency as published in the notice of availability. An EA does not require a public hearing for verbal comments. Following the mandated public comment period, the lead action agency responds to any comments, and certifies either a FONSI or a Notice of Intent (NOI) to prepare an EIS in its public environmental review record. The preparation of an EIS then generates a similar but more lengthy, involved and expensive process.

Environmental Impact Statement

The adequacy of an environmental impact statement (EIS) can be challenged in federal court. Major proposed projects have been blocked because of an agency's failure to prepare an acceptable EIS. One prominent example was the Westway landfill and highway development in and along the Hudson River in New York City. Another prominent case involved the Sierra Club suing the Nevada Department of Transportation over its denial of the club's request to issue a supplemental EIS addressing air emissions of particulate matter and hazardous air pollutants in the case of widening U.S. Route 95 through Las Vegas. The case reached the United States Court of Appeals for the Ninth Circuit, which led to construction on the highway being halted until the court's final decision. The case was settled prior to the court's final decision.

Several state governments that have adopted "little NEPAs," state laws imposing EIS requirements for particular state actions. Some of those state laws such as the California Environmental Quality Act refer to the required environmental impact study as an environmental impact report.

This variety of state requirements produces voluminous data not just upon impacts of individual projects, but also in insufficiently researched scientific domains. For example, in a seemingly routine *Environmental Impact Report* for the city of Monterey, California, information came to light that led to the official federal endangered species listing of Hickman's potentilla, a rare coastal wildflower.

Transboundary Application

Environmental threats do not respect national borders. International pollution can have detrimental effects on the atmosphere, oceans, rivers, aquifers, farmland, the weather and biodiversity. Global climate change is transnational. Specific pollution threats include acid rain, radioactive contamination, debris in outer space, stratospheric ozone depletion and toxic oil spills. The Chernobyl disaster, precipitated by a nuclear accident on April 26, 1986, is a stark reminder of the devastating effects of transboundary nuclear pollution.

Environmental protection is inherently a cross-border issue and has led to the creation of transnational regulation via multilateral and bilateral treaties. The United Nations Conference on the Human Environment (UNCHE or Stockholm Conference) held in Stockholm in 1972 and the United Nations Conference on the Environment and Development (UNCED or Rio Summit, Rio Conference, or Earth Summit) held in Rio de Janeiro in 1992 were key in the creation of about 1,000 international instruments that include at least some provisions related to the environment and its protection.

The United Nations Economic Commission for Europe's Convention on Environmental Impact Assessment in a Transboundary Context was negotiated to provide an international legal framework for transboundary EIA.

However, as there is no universal legislature or administration with a comprehensive mandate, most international treaties exist parallel to one another and are further developed without the benefit of consideration being given to potential conflicts with other agreements. There is also the issue of international enforcement. This has led to duplications and failures, in part due to an inability to enforce agreements. An example is the failure of many international fisheries regimes to restrict harvesting practises.

Criticism

As per Jay *et al.*, EIA is used as a decision aiding tool rather than decision making tool. There is growing dissent about them as their influence on decisions is limited. Improved training for practitioners, guidance on best practice and continuing research have all been proposed.

EIAs have been criticized for excessively limiting their scope in space and time. No accepted procedure exists for determining such boundaries. The boundary refers to 'the spatial and temporal boundary of the proposal's effects'. This boundary is determined by the applicant and the lead assessor, but in practice, almost all EIAs address only direct and immediate on-site effects.

Development causes both direct and indirect effects. Consumption of goods and services, production, use and disposal of building materials and machinery, additional land use for activities of manufacturing and services, mining and refining, etc., all have environmental impacts. The indirect effects of development can be much higher than the direct effects examined by an EIA. Proposals such as airports or shipyards cause wide-ranging national and international effects, which should be covered in EIAs.

Broadening the scope of EIA can benefit the conservation of threatened species. Instead of concentrating on the project site, some EIAs employed a habitat-based approach that focused on much broader relationships among humans and the environment. As a result, alternatives that reduce the negative effects to the population of whole species, rather than local subpopulations, can be assessed.

Thissen and Agusdinata have argued that little attention is given to the systematic identification and assessment of uncertainties in environmental studies which is critical in situations where uncertainty cannot be easily reduced by doing more research. In line with this, Maier et al. have concluded on the need to consider uncertainty at all stages of the decision-making process. In such a way decisions can be made with confidence or known uncertainty. These proposals are justified on data that shows that environmental assessments fail to predict accurately the impacts observed. Tenney et al. and Wood et al. have reported evidence of the intrinsic uncertainty attached to EIAs predictions from a number of case studies worldwide. The gathered evidence consisted of comparisons between predictions in EIAs and the impacts measured during, or following project implementation. In explaining this trend, Tenney et al. have highlighted major causes such as project changes, modelling errors, errors in data and assumptions taken and bias introduced by people in the projects analyzed. Cardenas and Halman provide a comprehensive review on the issues of uncertainty in environmental impact assessments.

Air Quality Index

Smog builds up under an inversion in Almaty, Kazakhstan resulting in a high AQI

An air quality index (AQI) is a number used by government agencies to communicate to the public how polluted the air currently is or how polluted it is forecast to become. As the AQI increases, an increasingly large percentage of the population is likely to experience increasingly severe adverse health effects. Different countries have their own air quality indices, corresponding to different national air quality standards. Some of these are the Air Quality Health Index (Canada), the Air Pollution Index (Malaysia), and the Pollutant Standards Index (Singapore).

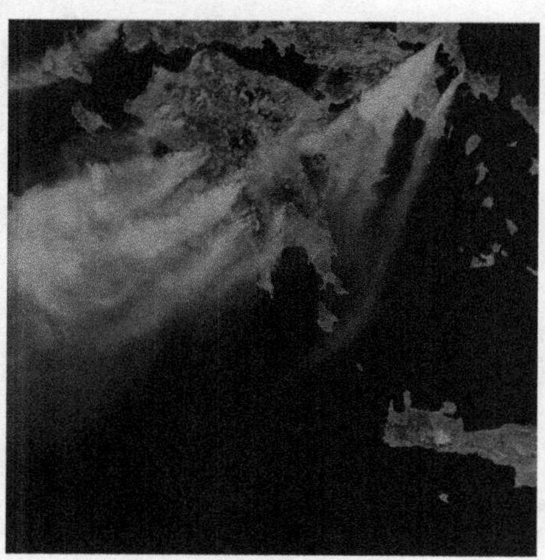

Wildfires give rise to an elevated AQI in parts of Greece

Definition and Usage

An air quality measurement station in Edinburgh, Scotland

Computation of the AQI requires an air pollutant concentration over a specified averaging period, obtained from an air monitor or model. Taken together, concentration and time represent the dose of the air pollutant. Health effects corresponding to a given dose are established by epidemiological research. Air pollutants vary in potency, and the function used to convert from air pollutant concentration to AQI varies by pollutant. Air quality index values are typically grouped into ranges. Each range is assigned a descriptor, a color code, and a standardized public health advisory.

The AQI can increase due to an increase of air emissions (for example, during rush hour traffic or when there is an upwind forest fire) or from a lack of dilution of air pollutants. Stagnant air, often caused by an anticyclone, temperature inversion, or low wind speeds lets air pollution remain in a local area, leading to high concentrations of pollutants, chemical reactions between air contaminants and hazy conditions.

Signboard in Gulfton, Houston indicating an ozone watch

On a day when the AQI is predicted to be elevated due to fine particle pollution, an agency or public health organization might:

- advise sensitive groups, such as the elderly, children, and those with respiratory or cardiovascular problems to avoid outdoor exertion.
- declare an "action day" to encourage voluntary measures to reduce air emissions, such as using public transportation.
- recommend the use of masks to keep fine particles from entering the lungs

During a period of very poor air quality, such as an air pollution episode, when the AQI indicates that acute exposure may cause significant harm to the public health, agencies may invoke emergency plans that allow them to order major emitters (such as coal burning industries) to curtail emissions until the hazardous conditions abate.

Most air contaminants do not have an associated AQI. Many countries monitor ground-level ozone, particulates, sulfur dioxide, carbon monoxide and nitrogen dioxide, and calculate air quality indices for these pollutants.

The definition of the AQI in a particular nation reflects the discourse surrounding the development of national air quality standards in that nation. A website allowing government agencies anywhere in the world to submit their real-time air monitoring data for display using a common definition of the air quality index has recently become available.

Indices by Location

Canada

Air quality in Canada has been reported for many years with provincial Air Quality Indices (AQIs). Significantly, AQI values reflect air quality management objectives, which are based on the lowest achievable emissions rate, and not exclusively concern for human health. The Air Quality Health Index or (AQHI) is a scale designed to help understand the impact of air quality on health. It is a health protection tool used to make decisions to reduce short-term exposure to air pollution by adjusting activity levels during increased levels of air pollution. The Air Quality Health Index also provides advice on how to improve air quality by proposing behavioural change to reduce the environmental footprint. This index pays particular attention to people who are sensitive to air pollution. It provides them with advice on how to protect their health during air quality levels associated with low, moderate, high and very high health risks.

The Air Quality Health Index provides a number from 1 to 10+ to indicate the level of health risk associated with local air quality. On occasion, when the amount of air pollution is abnormally high, the number may exceed 10. The AQHI provides a local air quality current value as well as a local air quality maximums forecast for today, tonight, and tomorrow, and provides associated health advice.

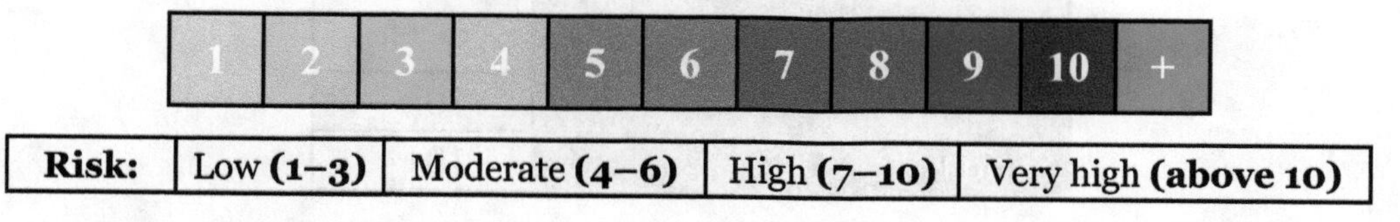

Health Risk	Air Quality Health Index	Health Messages	
		At Risk population	***General Population**
Low	**1–3**	**Enjoy** your usual outdoor activities.	**Ideal** air quality for outdoor activities
Moderate	**4–6**	**Consider reducing** or rescheduling strenuous activities outdoors if you are experiencing symptoms.	**No need to modify** your usual outdoor activities unless you experience symptoms such as coughing and throat irritation.
High	**7–10**	**Reduce** or reschedule strenuous activities outdoors. Children and the elderly should also take it easy.	**Consider reducing** or rescheduling strenuous activities outdoors if you experience symptoms such as coughing and throat irritation.
Very high	**Above 10**	**Avoid** strenuous activities outdoors. Children and the elderly should also avoid outdoor physical exertion.	**Reduce** or reschedule strenuous activities outdoors, especially if you experience symptoms such as coughing and throat irritation.

Hong Kong

On the 30th December 2013 Hong Kong replaced the Air Pollution Index with a new index called the *Air Quality Health Index*. This index is on a scale of 1 to 10+ and considers four air pollutants: ozone; nitrogen dioxide; sulphur dioxide and particulate matter (including PM10 and PM2.5). For any given hour the AQHI is calculated from the sum of the percentage excess risk of daily hospital admissions attributable to the 3-hour moving average concentrations of these four pollutants. The AQHIs are grouped into five AQHI health risk categories with health advice provided:

Health risk category	AQHI
Low	1
	2
	3
Medium	4
	5
	6
High	7
Very High	8
	9
	10
Serious	10+

Each of the health risk categories has advice with it. At the *low* and *moderate* levels the public are advised that they can continue normal activities. For the *high* category, children, the elderly and people with heart or respiratory illnesses are advising to reduce outdoor physical exertion. Above this (*very high* or *serious*) the general public are also advised to reduce or avoid outdoor physical exertion.

Mainland China

China's Ministry of Environmental Protection (MEP) is responsible for measuring the level of air pollution in China. As of 1 January 2013, MEP monitors daily pollution level in 163 of its major cities. The API level is based on the level of 6 atmospheric pollutants, namely sulfur dioxide (SO_2), nitrogen dioxide (NO_2), suspended particulates smaller than 10 μm in aerodynamic diameter (PM_{10}), suspended particulates smaller than 2.5 μm in aerodynamic diameter ($PM_{2.5}$), carbon monoxide (CO), and ozone (O_3) measured at the monitoring stations throughout each city.

AQI Mechanics

An individual score (IAQI) is assigned to the level of each pollutant and the final AQI is the highest of those 6 scores. The pollutants can be measured quite differently. $PM_{2.5}$ 、PM_{10} concentration are measured as average per 24h. SO_2, NO_2, O_3, CO are measured as average per hour. The final API value is calculated per hour according to a formula published by the MEP.

The scale for each pollutant is non-linear, as is the final AQI score. Thus an AQI of 100 does not mean twice the pollution of AQI at 50, nor does it mean twice as harmful. While an AQI of 50 from day 1 to 182 and AQI of 100 from day 183 to 365 does provide an annual average of 75, it does *not* mean the pollution is acceptable even if the benchmark of 100 is deemed safe. This is because the benchmark is a 24-hour target. The annual average must match against the annual target. It is entirely possible to have safe air every day of the year but still fail the annual pollution benchmark.

AQI and Health Implications (HJ 663-2012)

AQI	Air Pollution Level	Health Implications
0–50	Excellent	No health implications.
51–100	Good	Few hypersensitive individuals should reduce outdoor exercise.
101–150	Lightly Polluted	Slight irritations may occur, individuals with breathing or heart problems should reduce outdoor exercise.
151–200	Moderately Polluted	Slight irritations may occur, individuals with breathing or heart problems should reduce outdoor exercise.
201–300	Heavily Polluted	Healthy people will be noticeably affected. People with breathing or heart problems will experience reduced endurance in activities. These individuals and elders should remain indoors and restrict activities.
300+	Severely Polluted	Healthy people will experience reduced endurance in activities. There may be strong irritations and symptoms and may trigger other illnesses. Elders and the sick should remain indoors and avoid exercise. Healthy individuals should avoid outdoor activities.

India

The Minister for Environment, Forests & Climate Change Shri Prakash Javadekar launched The National Air Quality Index (AQI) in New Delhi on 17 September 2014 under the Swachh Bharat Abhiyan. It is outlined as ‘One Number- One Colour-One Description’ for the common man to judge the air quality within his vicinity. The index constitutes part of the Government’s mission to introduce the culture of cleanliness. Institutional and infrastructural measures are being undertaken in order to ensure that the mandate of cleanliness is fulfilled across the country and the Ministry of Environment, Forests & Climate Change proposed to discuss the issues concerned regarding quality of air with the Ministry of Human Resource Development in order to include this issue as part of the sensitisation programme in the course curriculum.

While the earlier measuring index was limited to three indicators, the current measurement index had been made quite comprehensive by the addition of five additional parameters. Under the current measurement of air quality there are 8 parameters . The initiatives undertaken by the Ministry recently aimed at balancing environment and conservation and development as air pollution has been a matter of environmental and health concerns, particularly in urban areas.

The Central Pollution Control Board along with State Pollution Control Boards has been operating National Air Monitoring Program (NAMP) covering 240 cities of the country having more than 342 monitoring stations. In addition, continuous monitoring systems that provide data on near real-time basis are also installed in a few cities. They provide information on air quality in public domain in simple linguistic terms that is easily understood by a common person. Air Quality Index (AQI) is one such tool for effective dissemination of air quality information to people. As such an Expert Group comprising medical professionals, air quality experts, academia, advocacy groups, and SPCBs was constituted and a technical study was awarded to IIT Kanpur. IIT Kanpur and the Expert Group recommended an AQI scheme in 2014.

There are six AQI categories, namely Good, Satisfactory, Moderately polluted, Poor, Very Poor, and Severe. The proposed AQI will consider eight pollutants (PM_{10}, $PM_{2.5}$, NO_2, SO_2, CO, O_3, NH_3, and Pb) for which short-term (up to 24-hourly averaging period) National Ambient Air Quality Standards are prescribed. Based on the measured ambient concentrations, corresponding standards and likely health impact, a sub-index is calculated for each of these pollutants. The worst sub-index reflects overall AQI. Associated likely health impacts for different AQI categories and pollutants have been also been suggested, with primary inputs from the medical expert members of the group. The AQI values and corresponding ambient concentrations (health breakpoints) as well as associated likely health impacts for the identified eight pollutants are as follows:

AQI Category, Pollutants and Health Breakpoints								
AQI Category (Range)	**PM_{10} (24hr)**	**$PM_{2.5}$ (24hr)**	**NO_2 (24hr)**	**O_3 (8hr)**	**CO (8hr)**	**SO_2 (24hr)**	**NH_3 (24hr)**	**Pb (24hr)**
Good (0-50)	0-50	0-30	0-40	0-50	0-1.0	0-40	0-200	0-0.5
Satisfactory (51-100)	51-100	31-60	41-80	51-100	1.1-2.0	41-80	201-400	0.5-1.0
Moderately polluted (101-200)	101-250	61-90	81-180	101-168	2.1-10	81-380	401-800	1.1-2.0
Poor (201-300)	251-350	91-120	181-280	169-208	10-17	381-800	801-1200	2.1-3.0
Very poor (301-400)	351-430	121-250	281-400	209-748	17-34	801-1600	1200-1800	3.1-3.5
Severe (401-500)	430+	250+	400+	748+	34+	1600+	1800+	3.5+

AQI	Associated Health Impacts
Good (0-50)	Minimal impact
Satisfactory (51-100)	May cause minor breathing discomfort to sensitive people.
Moderately polluted (101–200)	May cause breathing discomfort to people with lung disease such as asthma, and discomfort to people with heart disease, children and older adults.
Poor (201-300)	May cause breathing discomfort to people on prolonged exposure, and discomfort to people with heart disease.
Very poor (301-400)	May cause respiratory illness to the people on prolonged exposure. Effect may be more pronounced in people with lung and heart diseases.
Severe (401-500)	May cause respiratory impact even on healthy people, and serious health impacts on people with lung/heart disease. The health impacts may be experienced even during light physical activity.

Mexico

The air quality in Mexico City is reported in IMECAs. The IMECA is calculated using the measurements of average times of the chemicals ozone (O_3), sulphur dioxide (SO_2), nitrogen dioxide (NO_2), carbon monoxide (CO), particles smaller than 2.5 micrometers ($PM_{2.5}$), and particles smaller than 10 micrometers (PM_{10}).

Singapore

Singapore uses the Pollutant Standards Index to report on its air quality, with details of the calculation similar but not identical to that used in Malaysia and Hong Kong The PSI chart below is grouped by index values and descriptors, according to the National Environment Agency.

PSI	Descriptor	General Health Effects
0–50		None
51–100	Moderate	Few or none for the general population
101–200	Unhealthy	Mild aggravation of symptoms among susceptible persons i.e. those with underlying conditions such as chronic heart or lung ailments; transient symptoms of irritation e.g. eye irritation, sneezing or coughing in some of the healthy population.
201–300	Very Unhealthy	Moderate aggravation of symptoms and decreased tolerance in persons with heart or lung disease; more widespread symptoms of transient irritation in the healthy population.
301–400	Hazardous	Early onset of certain diseases in addition to significant aggravation of symptoms in susceptible persons; and decreased exercise tolerance in healthy persons.
Above 400	Hazardous	PSI levels above 400 may be life-threatening to ill and elderly persons. Healthy people may experience adverse symptoms that affect normal activity.

South Korea

The Ministry of Environment of South Korea uses the Comprehensive Air-quality Index (CAI) to describe the ambient air quality based on the health risks of air pollution. The index aims to help the public easily understand the air quality and protect people's health. The CAI is on a scale from 0 to 500, which is divided into six categories. The higher the CAI value, the greater the level of air pollution. Of values of the five air pollutants, the highest is the CAI value. The index also has associated health effects and a colour representation of the categories as shown below.

CAI	Description	Health Implications
0–50	Good	A level that will not impact patients suffering from diseases related to air pollution.
51–100	Moderate	A level that may have a meager impact on patients in case of chronic exposure.
101–150	Unhealthy for sensitive groups	A level that may have harmful impacts on patients and members of sensitive groups.
151–250	Unhealthy	A level that may have harmful impacts on patients and members of sensitive groups (children, aged or weak people), and also cause the general public unpleasant feelings.
251–500	Very unhealthy	A level that may have a serious impact on patients and members of sensitive groups in case of acute exposure.

The N Seoul Tower on Namsan Mountain in central Seoul, South Korea, is illuminated in blue, from sunset to 23:00 and 22:00 in winter, on days where the air quality in Seoul is 45 or less. During the spring of 2012, the Tower was lit up for 52 days, which is four days more than in 2011.

United Kingdom

The most commonly used air quality index in the UK is the *Daily Air Quality Index* recommended by the Committee on Medical Effects of Air Pollutants (COMEAP). This index has ten points, which are further grouped into 4 bands: low, moderate, high and very high. Each of the bands comes with advice for at-risk groups and the general population.

Air pollution banding	Value	Health messages for At-risk individuals	Health messages for General population
Low	**1–3**	Enjoy your usual outdoor activities.	Enjoy your usual outdoor activities.
Moderate	**4–6**	Adults and children with lung problems, and adults with heart problems, who experience symptoms, should consider reducing strenuous physical activity, particularly outdoors.	Enjoy your usual outdoor activities.

High	7–9	Adults and children with lung problems, and adults with heart problems, should reduce strenuous physical exertion, particularly outdoors, and particularly if they experience symptoms. People with asthma may find they need to use their reliever inhaler more often. Older people should also reduce physical exertion.	Anyone experiencing discomfort such as sore eyes, cough or sore throat should consider reducing activity, particularly outdoors.
Very High	10	Adults and children with lung problems, adults with heart problems, and older people, should avoid strenuous physical activity. People with asthma may find they need to use their reliever inhaler more often.	Reduce physical exertion, particularly outdoors, especially if you experience symptoms such as cough or sore throat.

The index is based on the concentrations of 5 pollutants. The index is calculated from the concentrations of the following pollutants: Ozone, Nitrogen Dioxide, Sulphur Dioxide, PM2.5 (particles with an aerodynamic diameter less than 2.5 μm) and PM10. The breakpoints between index values are defined for each pollutant separately and the overall index is defined as the maximum value of the index. Different averaging periods are used for different pollutants.

Index	Ozone, Running 8 hourly mean (μg/m³)	Nitrogen Dioxide, Hourly mean (μg/m³)	Sulphur Dioxide, 15 minute mean (μg/m³)	PM2.5 Particles, 24 hour mean (μg/m³)	PM10 Particles, 24 hour mean (μg/m³)
1	0-33	0-67	0-88	0-11	0-16
2	34-66	68-134	89-177	12-23	17-33
3	67-100	135-200	178-266	24-35	34-50
4	101-120	201-267	267-354	36-41	51-58
5	121-140	268-334	355-443	42-47	59-66
6	141-160	335-400	444-532	48-53	67-75
7	161-187	401-467	533-710	54-58	76-83
8	188-213	468-534	711-887	59-64	84-91
9	214-240	535-600	888-1064	65-70	92-100
10	≥ 241	≥ 601	≥ 1065	≥ 71	≥ 101

Europe

To present the air quality situation in European cities in a comparable and easily understandable way, all detailed measurements are transformed into a single relative figure: the Common Air Quality Index (or CAQI) Three different indices have been developed by Citeair to enable the comparison of three different time scale:.

- An hourly index, which describes the air quality today, based on hourly values and updated every hours,
- A daily index, which stands for the general air quality situation of yesterday, based on daily values and updated once a day,
- An annual index, which represents the city's general air quality conditions throughout the year and compare to European air quality norms. This index is based on the pollutants year average compare to annual limit values, and updated once a year.

However, the proposed indices and the supporting common web site www.airqualitynow.eu are designed to give a dynamic picture of the air quality situation in each city but not for compliance checking.

The Hourly and Daily Common Indices

These indices have 5 levels using a scale from 0 (very low) to > 100 (very high), it is a relative measure of the amount of air pollution. They are based on 3 pollutants of major concern in Europe: PM10, NO2, O3 and will be able to take into account to 3 additional pollutants (CO, PM2.5 and SO2) where data are also available.

The calculation of the index is based on a review of a number of existing air quality indices, and it reflects EU alert threshold levels or daily limit values as much as possible. In order to make cities more comparable, independent of the nature of their monitoring network two situations are defined:

- Background, representing the general situation of the given agglomeration (based on urban background monitoring sites),
- Roadside, being representative of city streets with a lot of traffic, (based on roadside monitoring stations)

The indices values are updated hourly (for those cities that supply hourly data) and yesterdays daily indices are presented.

Common air quality index legend:

Pollution	Index Value
Very low	0/25
Low	25/50
Medium	50/75
High	75/100
Very high	>100

The Common Annual Air Quality Index

The common annual air quality index provides a general overview of the air quality situation in a given city all the year through and regarding to the European norms.

It is also calculated both for background and traffic conditions but its principle of calculation is different from the hourly and daily indices. It is presented as a distance to a target index, this target being derived from the EU directives (annual air quality standards and objectives):

- If the index is higher than 1: for one or more pollutants the limit values are not met.
- If the index is below 1: on average the limit values are met.

The annual index is aimed at better taking into account long term exposure to air pollution based on distance to the target set by the EU annual norms, those norms being linked most of the time to recommendations and health protection set up by World Health Organisation.

United States

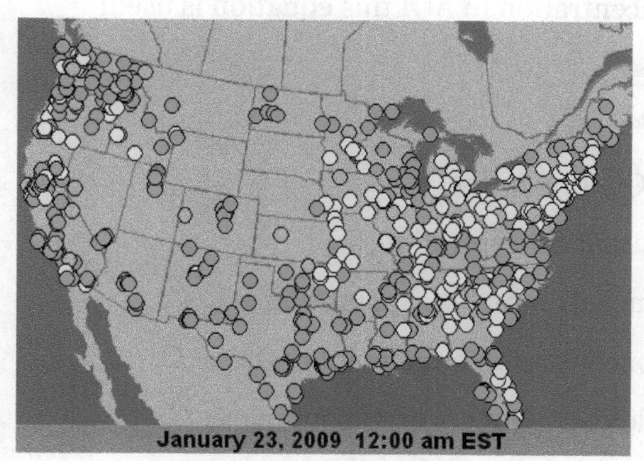

$PM_{2.5}$ 24-Hour AQI Loop, Courtesy US EPA

The United States Environmental Protection Agency (EPA) has developed an Air Quality Index that is used to report air quality. This AQI is divided into six categories indicating increasing levels of health concern. An AQI value over 300 represents hazardous air quality and below 50 the air quality is good.

Air Quality Index (AQI) Values	Levels of Health Concern	Colors
0 to 50	Good	Green
51 to 100	Moderate	Yellow
101 to 150	Unhealthy for Sensitive Groups	Orange

151 to 200	Unhealthy	Red
201 to 300	Very Unhealthy	Purple
301 to 500	Hazardous	Maroon

The AQI is based on the five "criteria" pollutants regulated under the Clean Air Act: ground-level ozone, particulate matter, carbon monoxide, sulfur dioxide, and nitrogen dioxide. The EPA has established National Ambient Air Quality Standards (NAAQS) for each of these pollutants in order to protect public health. An AQI value of 100 generally corresponds to the level of the NAAQS for the pollutant. The Clean Air Act (USA) (1990) requires EPA to review its National Ambient Air Quality Standards every five years to reflect evolving health effects information. The Air Quality Index is adjusted periodically to reflect these changes.

Computing the AQI

The air quality index is a piecewise linear function of the pollutant concentration. At the boundary between AQI categories, there is a discontinuous jump of one AQI unit. To convert from concentration to AQI this equation is used:

$$I = \frac{I_{high} - I_{low}}{C_{high} - C_{low}}(C - C_{low}) + I_{low}$$

where:

I = the (Air Quality) index,

C = the pollutant concentration,

C_{low} = the concentration breakpoint that is ≤ C,

C_{high} = the concentration breakpoint that is ≥ C,

I_{low} = the index breakpoint corresponding to C_{low},

I_{high} = the index breakpoint corresponding to C_{high}.

EPA's table of breakpoints is:

O_3 (ppb)	O_3 (ppb)	$PM_{2.5}$ (µg/m³)	PM_{10} (µg/m³)	CO (ppm)	SO_2 (ppb)	NO_2 (ppb)	AQI	AQI
C_{low} - C_{high} (avg)	C_{low} - C_{high} (avg)	C_{low} - C_{high} (avg)	C_{low} - C_{high} (avg)	C_{low} - C_{high} (avg)	C_{low} - C_{high} (avg)	C_{low} - C_{high} (avg)	I_{low} - I_{high}	**Category**
0-54 (8-hr)	-	0.0-12.0 (24-hr)	0-54 (24-hr)	0.0-4.4 (8-hr)	0-35 (1-hr)	0-53 (1-hr)	0-50	Good
55-70 (8-hr)	-	12.1-35.4 (24-hr)	55-154 (24-hr)	4.5-9.4 (8-hr)	36-75 (1-hr)	54-100 (1-hr)	51-100	Moderate

71-85 (8-hr)	125-164 (1-hr)	35.5-55.4 (24-hr)	155-254 (24-hr)	9.5-12.4 (8-hr)	76-185 (1-hr)	101-360 (1-hr)	101-150	Unhealthy for Sensitive Groups
86-105 (8-hr)	165-204 (1-hr)	55.5-150.4 (24-hr)	255-354 (24-hr)	12.5-15.4 (8-hr)	186-304 (1-hr)	361-649 (1-hr)	151-200	Unhealthy
106-200 (8-hr)	205-404 (1-hr)	150.5-250.4 (24-hr)	355-424 (24-hr)	15.5-30.4 (8-hr)	305-604 (24-hr)	650-1249 (1-hr)	201-300	Very Unhealthy
-	405-504 (1-hr)	250.5-350.4 (24-hr)	425-504 (24-hr)	30.5-40.4 (8-hr)	605-804 (24-hr)	1250-1649 (1-hr)	301-400	Hazardous
-	505-604 (1-hr)	350.5-500.4 (24-hr)	505-604 (24-hr)	40.5-50.4 (8-hr)	805-1004 (24-hr)	1650-2049 (1-hr)	401-500	

Suppose a monitor records a 24-hour average fine particle ($PM_{2.5}$) concentration of 12.0 micrograms per cubic meter. The equation above results in an AQI of:

$$\frac{50-0}{12.0-0}(12.0-0)+0=50,$$

corresponding to air quality in the "Good" range. To convert an air pollutant concentration to an AQI, EPA has developed a calculator.

If multiple pollutants are measured at a monitoring site, then the largest or "dominant" AQI value is reported for the location. The ozone AQI between 100 and 300 is computed by selecting the larger of the AQI calculated with a 1-hour ozone value and the AQI computed with the 8-hour ozone value.

8-hour ozone averages do not define AQI values greater than 300; AQI values of 301 or greater are calculated with 1-hour ozone concentrations. 1-hour SO_2 values do not define higher AQI values greater than 200. AQI values of 201 or greater are calculated with 24-hour SO_2 concentrations.

Real time monitoring data from continuous monitors are typically available as 1-hour averages. However, computation of the AQI for some pollutants requires averaging over multiple hours of data. (For example, calculation of the ozone AQI requires computation of an 8-hour average and computation of the $PM_{2.5}$ or PM_{10} AQI requires a 24-hour average.) To accurately reflect the current air quality, the multi-hour average used for the AQI computation should be centered on the current time, but as concentrations of future hours are unknown and are difficult to estimate accurately, EPA uses surrogate concentrations to estimate these multi-hour averages. For reporting the $PM_{2.5}$, PM_{10} and ozone air quality indices, this surrogate concentration is called the NowCast. The Nowcast is a particular type of weighted average that provides more weight to the most recent air quality data when air pollution levels are changing.

Public Availability of the AQI

Real time monitoring data and forecasts of air quality that are color-coded in terms of the air quality index are available from EPA's AirNow web site. Historical air monitoring data including AQI charts and maps are available at EPA's AirData website.

History of the AQI

The AQI made its debut in 1968, when the National Air Pollution Control Administration undertook an initiative to develop an air quality index and to apply the methodology to Metropolitan Statistical Areas. The impetus was to draw public attention to the issue of air pollution and indirectly push responsible local public officials to take action to control sources of pollution and enhance air quality within their jurisdictions.

Jack Fensterstock, the head of the National Inventory of Air Pollution Emissions and Control Branch, was tasked to lead the development of the methodology and to compile the air quality and emissions data necessary to test and calibrate resultant indices.

The initial iteration of the air quality index used standardized ambient pollutant concentrations to yield individual pollutant indices. These indices were then weighted and summed to form a single total air quality index. The overall methodology could use concentrations that are taken from ambient monitoring data or are predicted by means of a diffusion model. The concentrations were then converted into a standard statistical distribution with a preset mean and standard deviation. The resultant individual pollutant indices are assumed to be equally weighted, although values other than unity can be used. Likewise, the index can incorporate any number of pollutants although it was only used to combine SOx, CO, and TSP because of a lack of available data for other pollutants.

While the methodology was designed to be robust, the practical application for all metropolitan areas proved to be inconsistent due to the paucity of ambient air quality monitoring data, lack of agreement on weighting factors, and non-uniformity of air quality standards across geographical and political boundaries. Despite these issues, the publication of lists ranking metropolitan areas achieved the public policy objectives and led to the future development of improved indices and their routine application.

Freshwater Environmental Quality Parameters

Freshwater environmental quality parameters are the natural and man-made chemical, biological and microbiological characteristics of rivers, lakes and groundwaters, the ways they are measured and the ways that they change. The values or concentrations attributed to such parameters can be used to describe the pollution status of an environment, its biotic status or to predict the likelihood or otherwise of a particular organisms being present. Monitoring of environmental quality parameters is a key activity in managing the environment, restoring polluted environments and anticipating the effects of man-made changes on the environment.

Freshwater environmental quality parameters are those chemical, physical or biological parameters that can be used to characterise a freshwater body. Because almost all water bodies are dynamic in their composition, the relevant quality parameters are typically expressed as a range of expected concentrations.

Characterisation

The first step in understanding the chemistry of freshwater is to establish the relevant concentrations of the parameters of interest. Conventionally this is done by taking representative samples of the water for subsequent analysis in a laboratory . However, in-situ monitoring using hand-held analytical equipment or using bank-side monitoring stations are also used.

Sampling

Freshwaters are surprisingly difficult to sample because they are rarely homogeneous and their quality varies during the day and during the year. In addition the most representative sampling locations are often at a distance from the shore or bank increasing the logistic complexity.

Rivers

Filling a clean bottle with river water is a very simple task, but a single sample is only representative of that point along the river the sample was taken from and at that point in time. Understanding the chemistry of a whole river, or even a significant tributary, requires prior investigative work to understand how homogeneous or mixed the flow is and to determine if the quality changes during the course of a day and during the course of a year. Almost all natural rivers will have very significant patterns of change through the day and through the seasons. Many rivers also have a very large flow that is unseen. This flows through underlying gravel and sand layers and is called the hyporheic zone How much mixing there is between the hyporheic zone and the water in the open channel will depend on a variety of factors, some of which relate to flows leaving aquifers which may have been storing water for many years.

Ground-waters

Ground waters by their very nature are often very difficult to access to take a sample. As a consequence the majority of ground-water data comes from samples taken from springs, wells, water supply bore-holes and in natural caves. In recent decades as the need to understand ground water dynamics has increased, an increasing number or monitoring bore-holes have been drilled into aquifers

Lakes

Lakes and ponds can be very large and support a complex eco-system in which environmental parameters vary widely in all three physical dimensions and with time. Large lakes in the temperate zone often stratify in the warmer months into a warmer upper layers rich in oxygen and a colder lower layer with low oxygen levels. In the autumn, falling temperatures and occasional high winds result in the mixing of the two layers into a more homogeneous whole. When stratification occurs it not only affects oxygen levels but also many related parameters such as iron, phosphate and manganese which are all changed in their chemical form by change in the redox potential of the environment.

Lakes also receive waters, often from many different sources with varying qualities. Solids from stream inputs will typically settle near the mouth of the stream and depending on a variety of factors the incoming water may float over the surface of the lake, sink beneath the surface or rapidly mix with the lake water. All of these phenomena can skew the results of any environmental monitoring unless the process are well understood.

Mixing Zones

Where two rivers meet at a confluence there exists a mixing zone. A mixing zone may be very large and extend for many miles as in the case of the Mississippi and Missouri rivers in the United States and the River Clwyd and River Elwy in North Wales. In a mixing zone water chemistry may be very variable and can be difficult to predict. The chemical interactions are not just simple mixing but may be complicated by biological processes from submerged macrophytes and by water joining the channel from the hyporheic zone or from springs draining an aquifer.

Geological Inputs

The geology that underlies a river or lake has a major impact on its chemistry. A river flowing across very ancient precambrian schists is likely to have dissolved very little from the rocks and maybe similar to de-ionised water at least in the headwaters. Conversely a river flowing through chalk hills, and especially if its source is in the chalk, will have a high concentration of carbonates and bicarbonates of Calcium and possibly Magnesium.

As a river progresses along its course it may pass through a variety of geological types and it may have inputs from aquifers that do not appear on the surface anywhere in the locality.

Atmospheric Inputs

Oxygen is probably the most important chemical constituent of surface water chemistry, as all aerobic organisms require it for survival. It enters the water mostly via diffusion at the water-air interface. Oxygen's solubility in water decreases as water temperature increases. Fast, turbulent streams expose more of the water's surface area to the air and tend to have low temperatures and thus more oxygen than slow, backwaters. Oxygen is a by-product of photosynthesis, so systems with a high abundance of aquatic algae and plants may also have high concentrations of oxygen during the day. These levels can decrease significantly during the night when primary producers switch to respiration. Oxygen can be limiting if circulation between the surface and deeper layers is poor, if the activity of animals is very high, or if there is a large amount of organic decay occurring such as following Autumn leaf-fall.

Most other atmospheric inputs come from man-made or anthropogenic sources the most significant of which are the oxides of sulphur produced by burning sulphur rich fuels such as coal and oil which give rise to acid rain. The chemistry of sulphur oxides is complex both in the atmosphere and in river systems. However the effect on the overall chemistry is simple in that it reduces the pH of the water making it more acidic. The pH change is most marked in rivers with very low concentrations of dissolved salts as these cannot buffer the effects of the acid input. Rivers downstream of major industrial conurbations are also at greatest risk. In parts of Scandinavia and West Wales and Scotland many rivers became so acidic from oxides of sulphur that most fish life was destroyed and pHs as low as pH4 were recorded during critical weather conditions.

Anthropogenic Inputs

The majority of rivers on the planet and many lakes have received or are receiving inputs from human-kind's activities. In the industrialised world, many rivers have been very seriously polluted, at least during the 19th and the first half of the 20th centuries. Although in general there has been much improvement in the developed world, there is still a great deal of river pollution apparent on the planet.

Toxicity

In most environmental situations the presence or absence of an organism is determined by a complex web of interactions only some of which will be related to measurable chemical or biological parameters. Flow rate, turbulence, inter and intra specific competition, feeding behaviour, disease, parasatism, commensalism and symbiosis are just a few of the pressures and opportunities facing any organism or population. Most chemical constituents favour some organisms and are less favourable to others. However, there are some cases where a chemical constituent exerts a toxic effect. i.e. where the concentration can kill or severely inhibit the normal functioning of the organism. Where a toxic effect has been demonstrated this may be noted in the sections below dealing with the individual parameters.

Chemical Constituents

Colour and Turbidity

Often it is the colour of freshwater or how clear or hazy the water is that is the most obvious visual characteristic. Unfortunately neither colour nor turbidity are strong indicators of the overall chemical composition of water. However both colour and turbidity reduce the amount of light penetrating the water and can have significant impact on algae and macrophytes. Some algae in particular are highly dependent on water with low colour and turbidity

Many rivers draining high moor-lands overlain by peat have a very deep yellow brown colour caused by dissolved humic acids.

Organic Constituents

One of the principal sources of elevated concentrations of organic chemical constituents is from treated sewage.

Dissolved organic material is most commonly measured using either the Biochemical oxygen demand (BOD) test or the Chemical oxygen demand (COD) test. Organic constituents are significant in river chemistry for the effect that they have on dissolved oxygen concentration and for the impact that individual organic species may have directly on aquatic biota.

Any organic and degradable material consumes oxygen as it decomposes. Where organic concentrations are significantly elevated the effects on oxygen concentrations can be significant and as conditions get extreme the river bed may become anoxic.

Some organic constituents such as synthetic hormones, pesticides, phthalates have direct metabolic effects on aquatic biota and even on humans drinking water taken from the river. Understanding such constituents and how they can be identified and quantified is becoming of increasing importance in the understanding of freshwater chemistry.

Metals

A wide range of metals may be found in rivers from natural sources where metal ores are present in the rocks over which the river flows or in the aquifers feeding water into the river. However many rivers have an increased load of metals because of industrial activities which include mining and quarrying and the processing and use of metals.

Iron

Iron, usually as Fe^{+++} is a common constituent of river waters at very low levels. Higher iron concentrations in acidic springs or an anoxic hyporheic zone may cause visible orange/brown staining or semi-gelatinous precipitates of dense orange iron bacterial floc carpeting the river bed. Such conditions are very deleterious to most organisms and can cause serious damage in a river system.

Coal mining is also a very significant source of Iron both in mine-waters and from stocking yards of coal and from coal processing. Long abandoned mines can be a highly intractable source of high concentrations of Iron. Low levels of iron are common in spring waters emanating from deep-seated aquifers and maybe regarding as health giving springs. Such springs are commonly called Chalybeate springs and have given rise to a number of Spa towns in Europe and the United States.

Zinc

Zinc is normally associated with metal mining, especially Lead and Silver mining but is also a component pollutant associated with a variety of other metal mining activities and with Coal mining. Zinc is toxic at relatively low concentrations to many aquatic organisms. *Microregma* starts to show a toxic reaction at concentrations as low as 0.33 mg/l

Heavy Metals

Lead and silver in river waters are commonly found together and associated with lead mining. Impacts from very old mines can be very long-lived. In the River Ystwyth in Wales for example, the effects of silver and lead mining in the 17th and 18th centuries in the headwaters still causes unacceptably high levels of Zinc and Lead in the river water right down to its confluence with the sea. Silver is very toxic even at very low concentrations but leaves no visible evidence of its contamination.

Lead is also highly toxic to freshwater organisms and to humans if the water is used as drinking water. As with Silver, Lead pollution is not visible to the naked eye. The River Rheidol in west Wales had a major series of lead mines in its headwaters until the end of the 19th century and its mine discharges and waste tips remain to this day. In 1919 - 1921 only 14 species of invertebrates were found in the lower Rheidol when Lead concentrations were between 0.2ppm and 0.5ppm. By 1932 the lead concentration had reduced to 0.02ppm to 0.1ppm because of the abandonment of mining and, at those concentrations, the bottom fauna had stabilized to 103 species including three leeches.

Coal mining is also a very significant source of metals, especially Iron, Zinc and Nickel particularly where the coal is rich if pyrites which oxidises on contact with the air producing a very acidic leachate which is able to dissolve metals from the coal.

Significant levels of copper are unusual in rivers and where it does it occur the source is most likely to be mining activities, coal stocking, or pig farming. Rarely elevated levels may be of geological origin. Copper is acutely toxic to many freshwater organisms, especially algae, at very low concentrations and significant concentration in river water may have serious adverse effects on the local ecology.

Nitrogen

Nitrogenous compounds have a variety of sources including washout of oxides of nitrogen from the atmosphere, some geological inputs and some from macrophyte and algal nitrogen fixation. However, for many rivers in the proximity of humans, the largest input is from sewage whether treated or untreated. The nitrogen derives from breakdown products of proteins found in urine and faeces. These products, being very soluble, often pass through sewage treatment process and are discharged into rivers as a component of sewage treatment effluent. Nitrogen may be in the form of nitrate, nitrite, ammonia or ammonium salts or what is termed albuminoid nitrogen or nitrogen still within an organic proteinoid molecule.

The differing forms of nitrogen are relatively stable in most river systems with nitrite slowly transforming into nitrate in well oxygenated rivers and ammonia transforming into nitrite/ nitrate. However, the process are slow in cool rivers and reduction in concentration may more often be attributed to simple dilution. All forms of nitrogen are taken up by macrophytes and algae and elevated levels of nitrogen are often associated with overgrowths of plants or eutrophication. These can have the effect of blocking channels and inhibiting navigation. However, ecologically, the more significant effect is on dissolved oxygen concentrations which may become super-saturated during daylight due to plant photosynthesis but then drop to very low levels during darkness as plant respiration uses up the dissolved oxygen. Coupled with the release of oxygen in photosynthesis is the creation of bi-carbonate ions which cause a steep rise in pH and this is matched in darkness as carbon dioxide is released through respiration which substantially lowers the pH. Thus high levels of nitrogenous compounds tends to lead to eutrophication with extreme variations in parameters which in turn can substantially degrade the ecological worth of the watercourse.

Ammonium ions also have a toxic effect, especially on fish. The toxicity of ammonia is dependent on both pH and temperature and an added complexity is the buffering effect of the blood/water interface across the gill membrane which masks any additional toxicity over about pH 8.0. The management of river chemistry to avoid ecological damage is particularly difficult in the case of ammonia as a wide range of potential scenarios of concentration, pH and temperature have to be considered and the diurnal pH fluctuation caused by photosynthesis considered. On warm summer days with high-bi-carbonate concentrations unexpectedly toxic conditions can be created.

Phosphorus

Phosphorus compounds are usually found as relatively insoluble phosphates in river water and, except in some exceptional circumstances, their origin is agriculture or human sewage. Phosphorus can encourage excessive growths of plants and algae and contribute to eutrophication. If a river discharges into a lake or reservoir phosphate can be mobilised year after year by natural processes. In the summer time, lakes stratify so that warm oxygen rich water floats on top of cold oxygen poor water. In the warm upper layers - the epilimnion-plants consume the available phosphate. As the plants die in the late summer they fall into the cool water layers underneath - the hypolimnion - and decompose. During winter turn-over, when a lake becomes fully mixed through the action of winds on a cooling body of water - the phosphates are spread throughout the lake again to feed a new generation of plants. This process is one of the principal causes of persistent algal blooms at some lakes.

Arsenic

Geological deposits of arsenic may be released into rivers where deep ground-waters are exploited as in parts of Pakistan. Many metalloid ores such as lead, gold and copper contain traces of arsenic and poorly stored tailings may result in arsenic entering the hydrological cycle.

Solids

Inert solids are produced in all montane rivers as the energy of the water helps grind away rocks into gravel, sand and finer material. Much of this settles very quickly and provides an important substrate for many aquatic organisms. Many salmonid fish require beds of gravel and sand in which to lay their eggs. Many other types of solids from agriculture, mining, quarrying, urban run-off and sewage may block-out sunlight from the river and may block interstices in gravel beds making them useless for spawning and supporting insect life.

Bacterial, Viral and Parasite Inputs

Both agriculture and sewage treatment produce inputs into rivers with very high concentrations of bateria and viruses including a wide range of pathogenic organisms. Even in areas with little human activity significant levels of bacteria and viruses can be detected originating from fish and aquatic mammals and from animals grazing near rivers such as deer. Upland waters draining areas frequented by sheep, goats or deer may also harbour a variety of opportunistic human parasites such as liver fluke. Consequently, there are very few rivers from which the water is safe to drink without some form of sterilisation or disinfection. In rivers used for contact recreation such as swimming, safe levels of bacteria and viruses can be established based on risk assessment.

Under certain conditions bacteria can colonise freshwaters occasionally making large rafts of filamentous mats known as *sewage fungus* – usually *Sphaerotilus natans*. The presence of such organisms is almost always an indicator of extreme organic pollution and would be expected to be matched with low dissolved oxygen concentrations and high BOD vales.

E. coli bacteria have been commonly found in recreational waters and their presence is used to indicate the presence of recent fecal contamination, but E. coli presence may not be indicative of human waste. E. coli are harbored in all warm-blooded animals: birds and mammals alike. E. coli bacteria have also been found in fish and turtles. Sand also harbors E. coli bacteria and some strains of E. coli have become naturalized. Some geographic areas may support unique populations of E. coli and conversely, some E. coli strains are cosmopolitan .

pH

pH in rivers is affected by the geology of the water source, atmospheric inputs and a range of other chemical contaminants. pH is only likely to become an issue on very poorly buffered upland rivers where atmospheric sulphur and nitrogen oxides may very significantly depress the pH as low as pH4 or in eutrophic alkaline rivers where photosynthetic bi-carbonate ion production in photosynthesis may drive the pH up above pH10

Water Quality

A rosette sampler is used to collect water samples in deep water, such as the Great Lakes or oceans, for water quality testing.

Water quality refers to the chemical, physical, biological, and radiological characteristics of water. It is a measure of the condition of water relative to the requirements of one or more biotic species and or to any human need or purpose. It is most frequently used by reference to a set of standards against which compliance can be assessed. The most common standards used to assess water quality relate to health of ecosystems, safety of human contact, and drinking water.

Standards

In the setting of standards, agencies make political and technical/scientific decisions about how the water will be used. In the case of natural water bodies, they also make some reasonable estimate of pristine conditions. Different uses raise different concerns and therefore different standards are considered. Natural water bodies will vary in response to environmental conditions. Environmental scientists work to understand how these systems function, which in turn helps to identify the sources and fates of contaminants. Environmental lawyers and policymakers work to define legislation with the intention that water is maintained at an appropriate quality for its identified use.

The vast majority of surface water on the planet is neither potable nor toxic. This remains true when seawater in the oceans (which is too salty to drink) is not counted. Another general perception of *water quality* is that of a simple property that tells whether water is polluted or not. In fact, water quality is a complex subject, in part because water is a complex medium intrinsically tied to the ecology of the Earth. Industrial and commercial activities (e.g. manufacturing, mining, construction, transport) are a major cause of water pollution as are runoff from agricultural areas, urban runoff and discharge of treated and untreated sewage.

Categories

The parameters for water quality are determined by the intended use. Work in the area of water quality tends to be focused on water that is treated for human consumption, industrial use, or in the environment.

Human Consumption

Contaminants that may be in untreated water include microorganisms such as viruses, protozoa and bacteria; inorganic contaminants such as salts and metals; organic chemical contaminants from industrial processes and petroleum use; pesticides and herbicides; and radioactive contaminants. Water quality depends on the local geology and ecosystem, as well as human uses such as sewage dispersion, industrial pollution, use of water bodies as a heat sink, and overuse (which may lower the level of the water).

The United States Environmental Protection Agency (EPA) limits the amounts of certain contaminants in tap water provided by US public water systems. The Safe Drinking Water Act authorizes EPA to issue two types of standards: *primary standards* regulate substances that potentially affect human health, and *secondary standards* prescribe aesthetic qualities, those that affect taste, odor, or appearance. The U.S. Food and Drug Administration (FDA) regulations establish limits for contaminants in bottled water that must provide the same protection for public health. Drinking water, including bottled water, may reasonably be expected to contain at least small amounts of some contaminants. The presence of these contaminants does not necessarily indicate that the water poses a health risk.

In urbanized areas around the world, water purification technology is used in municipal water systems to remove contaminants from the source water (surface water or groundwater) before it is distributed to homes, businesses, schools and other recipients. Water drawn directly from a stream, lake, or aquifer and that has no treatment will be of uncertain quality.

Industrial and Domestic Use

Dissolved minerals may affect suitability of water for a range of industrial and domestic purposes. The most familiar of these is probably the presence of ions of calcium and magnesium which interfere with the cleaning action of soap, and can form hard sulfate and soft carbonate deposits in water heaters or boilers. Hard water may be softened to remove these ions. The softening process often substitutes sodium cations. Hard water may be preferable to soft water for human consumption, since health problems have been associated with excess sodium and with calcium and magnesium deficiencies. Softening decreases nutrition and may increase cleaning effectiveness. Various industries' wastes and effluents can also pollute the water quality in receiving bodies of water.

Environmental Water Quality

Urban runoff discharging to coastal waters

Satirical cartoon by William Heath, showing a woman observing monsters in a drop of London water (at the time of the *Commission on the London Water Supply* report, 1828)

Environmental water quality, also called ambient water quality, relates to water bodies such as lakes, rivers, and oceans. Water quality standards for surface waters vary significantly due to different environmental conditions, ecosystems, and intended human uses. Toxic substances and high populations of certain microorganisms can present a health hazard for non-drinking purposes such as irrigation, swimming, fishing, rafting, boating, and industrial uses. These conditions may also affect wildlife, which use the water for drinking or as a habitat. Modern water quality laws generally specify protection of fisheries and recreational use and require, as a minimum, retention of current quality standards.

There is some desire among the public to return water bodies to pristine, or pre-industrial conditions. Most current environmental laws focus on the designation of particular uses of a water body. In some countries these designations allow for some water contamination as long as the particular type of contamination is not harmful to the designated uses. Given the landscape changes (e.g., land development, urbanization, clearcutting in forested areas) in the watersheds of many freshwater bodies, returning to pristine conditions would be a significant challenge. In these cases, environmental scientists focus on achieving goals for maintaining healthy ecosystems and may concentrate on the protection of populations of endangered species and protecting human health.

Sampling and Measurement

The complexity of water quality as a subject is reflected in the many types of measurements of water quality indicators. The most accurate measurements of water quality are made on-site, because water exists in equilibrium with its surroundings. Measurements commonly made on-site and in direct contact with the water source in question include temperature, pH, dissolved oxygen, conductivity, oxygen reduction potential (ORP), turbidity, and Secchi disk depth.

Sample Collection

An automated sampling station installed along the East Branch Milwaukee River, New Fane, Wisconsin. The cover of the 24-bottle autosampler (center) is partially raised, showing the sample bottles inside. The autosampler was programmed to collect samples at time intervals, or proportionate to flow over a specified period. The data logger (white cabinet) recorded temperature, specific conductance, and dissolved oxygen levels.

More complex measurements are often made in a laboratory requiring a water sample to be collected, preserved, transported, and analyzed at another location. The process of water sampling introduces two significant problems. The first problem is the extent to which the sample may be representative of the water source of interest. Many water sources vary with time and with location. The measurement of interest may vary seasonally or from day to night or in response to some activity of man or natural populations of aquatic plants and animals. The measurement of interest may vary with distances from the water boundary with overlying atmosphere and underlying or confining soil. The sampler must determine if a single time and location meets the needs of the investigation, or if the water use of interest can be satisfactorily assessed by averaged values with time and/or location, or if critical maxima and minima require individual measurements over a range of times, locations and/or events. The sample collection procedure must assure correct weighting of individual sampling times and locations where averaging is appropriate. Where critical maximum or minimum values exist, statistical methods must be applied to observed variation to determine an adequate number of samples to assess probability of exceeding those critical values.

The second problem occurs as the sample is removed from the water source and begins to establish chemical equilibrium with its new surroundings - the sample container. Sample containers must be made of materials with minimal reactivity with substances to be measured; and pre-cleaning of sample containers is important. The water sample may dissolve part of the sample container and any residue on that container, or chemicals dissolved in the water sample may sorb onto the sample container and remain there when the water is poured out for analysis. Similar physical and chemical interactions may take place with any pumps, piping, or intermediate devices used to transfer the water sample into the sample container. Water collected from depths below the surface will normally be held at the reduced pressure of the atmosphere; so gas dissolved in the water may escape into unfilled space at the top of the container. Atmospheric gas present in that air space may also dissolve into the water sample. Other chemical reaction equilibria may change if the water sample changes temperature. Finely divided solid particles formerly suspended by water turbulence may settle to the bottom of the sample container, or a solid phase may form from biological growth or chemical precipitation. Microorganisms within the water sample may biochemically alter concentrations of oxygen, carbon dioxide, and organic compounds. Changing carbon dioxide concentrations may alter pH and change solubility of chemicals of interest. These problems are of special concern during measurement of chemicals assumed to be significant at very low concentrations.

Sample preservation may partially resolve the second problem. A common procedure is keeping samples cold to slow the rate of chemical reactions and phase change, and analyzing the sample as soon as possible; but this merely minimizes the changes rather than preventing them. A useful procedure for determining influence of sample containers during delay between sample collection and analysis involves preparation for two artificial samples in advance of the sampling event. One sample container is filled with water known from previous analysis to contain no detectable amount of the chemical of interest. This sample, called a "blank," is opened for exposure to the atmosphere when the sample of interest is collected, then resealed and transported to the laboratory with the sample for analysis to determine if sample holding procedures introduced any measurable amount of the chemical of interest. The second artificial sample is collected with the sample of interest, but then "spiked" with a measured additional amount of the chemical of interest at the time of collection. The blank and spiked samples are carried with the sample of interest and analyzed by the same methods at the same times to determine any changes indicating gains or losses during the elapsed time between collection and analysis.

Testing in Response to Natural Disasters and Other Emergencies

Inevitably after events such as earthquakes and tsunamis, there is an immediate response by the aid agencies as relief operations get underway to try and restore basic infrastructure and provide the basic fundamental items that are necessary for survival and subsequent recovery. Access to clean drinking water and adequate sanitation is a priority at times like this. The threat of disease increases hugely due to the large numbers of people living close together, often in squalid conditions, and without proper sanitation.

After a natural disaster, as far as water quality testing is concerned there are widespread views on the best course of action to take and a variety of methods can be employed. The key basic water quality parameters that need to be addressed in an emergency are bacteriological indicators of fecal contamination, free chlorine residual, pH, turbidity and possibly conductivity/total dissolved solids. There are a number of portable water test kits on the market widely used by aid and relief agencies for carrying out such testing.

After major natural disasters, a considerable length of time might pass before water quality returns to pre-disaster levels. For example, following the 2004 Indian Ocean Tsunami the Colombo-based International Water Management Institute (IWMI) monitored the effects of saltwater and concluded that the wells recovered to pre-tsunami drinking water quality one and a half years after the event. IWMI developed protocols for cleaning wells contaminated by saltwater; these were subsequently officially endorsed by the World Health Organization as part of its series of Emergency Guidelines.

Chemical Analysis

A gas chromatograph-mass spectrometer measures pesticides and other organic pollutants

The simplest methods of chemical analysis are those measuring chemical elements without respect to their form. Elemental analysis for oxygen, as an example, would indicate a concentration of 890,000 milligrams per litre (mg/L) of water sample because water is made of oxygen. The method selected to measure dissolved oxygen should differentiate between diatomic oxygen and oxygen combined with other elements. The comparative simplicity of elemental analysis has produced a large amount of sample data and water quality criteria for elements sometimes identified as heavy metals.

Water analysis for heavy metals must consider soil particles suspended in the water sample. These suspended soil particles may contain measurable amounts of metal. Although the particles are not dissolved in the water, they may be consumed by people drinking the water. Adding acid to a water sample to prevent loss of dissolved metals onto the sample container may dissolve more metals from suspended soil particles. Filtration of soil particles from the water sample before acid addition, however, may cause loss of dissolved metals onto the filter. The complexities of differentiating similar organic molecules are even more challenging.

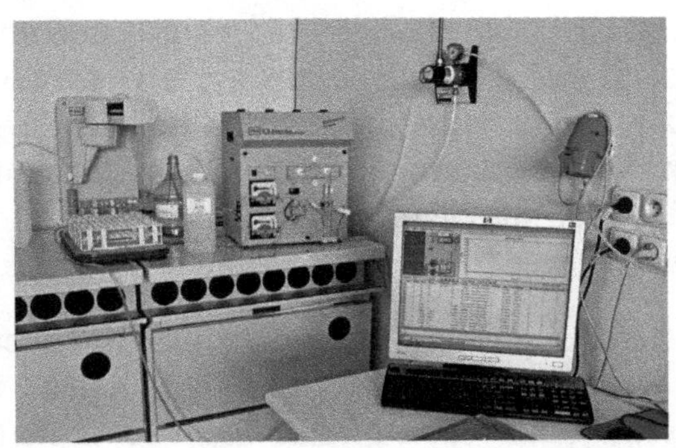

Atomic fluorescence spectroscopy is used to measure mercury and other heavy metals

Making these complex measurements can be expensive. Because direct measurements of water quality can be expensive, ongoing monitoring programs are typically conducted by government agencies. However, there are local volunteer programs and resources available for some general assessment. Tools available to the general public include on-site test kits, commonly used for home fish tanks, and biological assessment procedures.

Real-time Monitoring

Although water quality is usually sampled and analyzed at laboratories, nowadays, citizens demand real-time information about the water they are drinking. During the last years, several companies are deploying worldwide real-time remote monitoring systems for measuring water pH, turbidity or dissolved oxygen levels.

Drinking Water Indicators

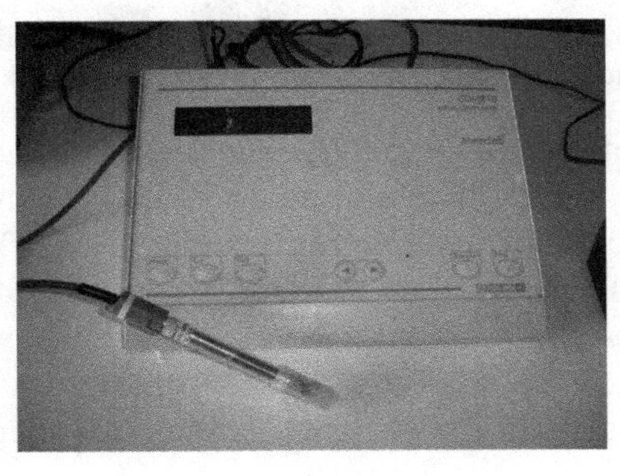

An electrical conductivity meter is used to measure total dissolved solids

The following is a list of indicators often measured by situational category:

- Alkalinity
- Color of water
- pH
- Taste and odor (geosmin, 2-Methylisoborneol (MIB), etc.)
- Dissolved metals and salts (sodium, chloride, potassium, calcium, manganese, magnesium)
- Microorganisms such as fecal coliform bacteria (*Escherichia coli*), Cryptosporidium, and Giardia lamblia
- Dissolved metals and metalloids (lead, mercury, arsenic, etc.)
- Dissolved organics: colored dissolved organic matter (CDOM), dissolved organic carbon (DOC)
- Radon
- Heavy metals
- Pharmaceuticals
- Hormone analogs

Environmental Indicators

Physical indicators	
• Water Temperature • Specifics Conductance or EC, Electrical Conductance, Conductivity • Total suspended solids (TSS) • Transparency or Turbidity	• Total dissolved solids (TDS) • Odour of water • Color of water • Taste of water
Chemical indicators	
• pH • Biochemical oxygen demand (BOD) • Chemical oxygen demand (COD) • Dissolved oxygen (DO) • Total hardness (TH)	• Heavy metals • Nitrate • Orthophosphates • Pesticides • Surfactants
Biological indicators	
• Ephemeroptera • Plecoptera • Mollusca • Trichoptera	• *Escherichia coli* (E. coli) • Coliform bacteria

Biological monitoring metrics have been developed in many places, and one widely used measure is the presence and abundance of members of the insect orders Ephemeroptera, Plecoptera and Trichoptera. (Common names are, respectively, Mayfly, Stonefly and Caddisfly.) EPT indexes will naturally vary from region to region, but generally, within a region, the greater the number of taxa from these orders, the better the water quality. Organisations in the United States, such as EPA offer guidance on developing a monitoring program and identifying members of these and other aquatic insect orders.

Individuals interested in monitoring water quality who cannot afford or manage lab scale analysis can also use biological indicators to get a general reading of water quality. One example is the IOWATER volunteer water monitoring program, which includes a benthic macroinvertebrate indicator key.

Bivalve molluscs are largely used as bioindicators to monitor the health of aquatic environments in both fresh water and the marine environments. Their population status or structure, physiology, behaviour or the level of contamination with elements or compounds can indicate the state of contamination status of the ecosystem. They are particularly useful since they are sessile so that they are representative of the environment where they are sampled or placed. A typical project is the Mussel Watch Programme, but today they are used worldwide.

The Southern African Scoring System (SASS) method is a biological water quality monitoring system based on the presence of benthic macroinvertebrates. The SASS aquatic biomonitoring tool has been refined over the past 30 years and is now on the fifth version (SASS5) which has been specifically modified in accordance with international standards, namely the ISO/IEC 17025 protocol. The SASS5 method is used by the South African Department of Water Affairs as a standard method for River Health Assessment, which feeds the national River Health Programme and the national Rivers Database.

International

- The World Health Organisation (WHO) has published guidelines for drinking-water quality (GDWQ) in 2011.
- The International Organization for Standardization (ISO) published[*when?*] regulation of water quality in the section of ICS 13.060, ranging from water sampling, drinking water, industrial class water, sewage, and examination of water for chemical, physical or biological properties. ICS 91.140.60 covers the standards of water supply systems.

National Specifications for Ambient Water and drinking water

European Union

The water policy of the European Union is primarily codified in three directives:

- Directive on Urban Waste Water Treatment (91/271/EEC) of 21 May 1991 concerning discharges of municipal and some industrial wastewaters;
- The Drinking Water Directive (98/83/EC) of 3 November 1998 concerning potable water quality;
- Water Framework Directive (2000/60/EC) of 23 October 2000 concerning water resources management.

India

- Indian Council of Medical Research (ICMR) Standards for Drinking Water.

South Africa

Water quality guidelines for South Africa are grouped according to potential user types (e.g. domestic, industrial) in the 1996 Water Quality Guidelines. Drinking water quality is subject to the South African National Standard (SANS) 241 Drinking Water Specification.

United Kingdom

In England and Wales acceptable levels for drinking water supply are listed in the "Water Supply (Water Quality) Regulations 2000."

United States

In the United States, Water Quality Standards are defined by state agencies for various water bodies, guided by the desired uses for the water body (e.g., fish habitat, drinking water supply, recreational use). The Clean Water Act (CWA) requires each governing jurisdiction (states, territories, and covered tribal entities) to submit a set of biennial reports on the quality of water in their area. These reports are known as the 303(d) and 305(b) reports, named for their respective CWA provisions, and are submitted to, and approved by, EPA. These reports are completed by the governing jurisdiction, typically a state environmental agency. EPA recommends that each state submit a single "Integrated Report" comprising its list of impaired waters and the status of all water bodies in the state. The *National Water Quality Inventory Report to Congress* is a general report on water quality, providing overall information about the number of miles of streams and rivers and their aggregate condition. The CWA requires states to adopt standards for each of the possible designated uses that they assign to their waters. Should evidence suggest or document that a stream, river or lake has failed to meet the water quality criteria for one or more of its designated uses, it is placed on a list of impaired waters. Once a state has placed a water body on this list, it must develop a management plan establishing Total Maximum Daily Loads (TMDLs) for the pollutant(s) impairing the use of the water. These TMDLs establish the reductions needed to fully support the designated uses.

Drinking water standards, which are applicable to public water systems, are issued by EPA under the Safe Drinking Water Act.

Oxygen Saturation

Measuring the dissolved oxygen through a multi-parameter photometer

Oxygen saturation (symbol SO_2) is a relative measure of the amount of oxygen that is dissolved or carried in a given medium. It can be measured with a dissolved oxygen probe such as an oxygen sensor or an optode in liquid media, usually water. The standard unit of oxygen saturation is percent (%).

Oxygen saturation can be measured regionally and noninvasively. Arterial oxygen saturation (SaO_2) is commonly measured using pulse oximetry. Tissue saturation at peripheral scale can be measured using NIRS. This technique can be applied on both muscle and brain.

In Medicine

In medicine, oxygen saturation refers to *oxygenation*, or when oxygen molecules (O2) enter the tissues of the body. In this case blood is oxygenated in the lungs, where oxygen molecules travel from the air and into the blood. Oxygen saturation ((O2) sats) is a measure the percentage of hemoglobin binding sites in the bloodstream occupied by oxygen. Fish, invertebrates, plants, and aerobic bacteria all require oxygen for respiration.

In Environmental Science

In aquatic environments, oxygen saturation is a ratio of the concentration of dissolved oxygen (O_2) in the water to the maximum amount of oxygen that will dissolve in the water at that temperature and pressure under stable equilibrium. Well-aerated water (such as a fast-moving stream) without oxygen producers or consumers is 100 % saturated.

It is possible for stagnant water to become somewhat supersaturated with oxygen (i.e. reach more than 100 % saturation) either because of the presence of photosynthetic aquatic oxygen producers or because of a slow equilibration after a change of atmospheric conditions. Stagnant water in the presence of decaying matter will typically have an oxygen concentration much less than 100 %.

Environmental oxygenation can be important to the sustainability of a particular ecosystem. Refer to (for a table of maximum equilibrium dissolved oxygen concentration versus temperature at atmospheric pressure. The optimal levels in an estuary for dissolved oxygen is higher than 6 ppm. Insufficient oxygen (environmental hypoxia), often caused by the decomposition of organic matter and/or nutrient pollution, may occur in bodies of water such as ponds and rivers, tending to suppress the presence of aerobic organisms such as fish. Deoxygenation increases the relative population of anaerobic organisms such as plants and some bacteria, resulting in fish kills and other adverse events. The net effect is to alter the balance of nature by increasing the concentration of anaerobic over aerobic species.

Chemical Oxygen Demand

In environmental chemistry, the chemical oxygen demand (COD) test is commonly used to indirectly measure the amount of organic compounds in water. Most applications of COD determine the amount of organic pollutants found in surface water (e.g. lakes and rivers) or wastewater, making COD a useful measure of water quality. It is expressed in milligrams per liter (mg/L), which indicates the mass of oxygen consumed per liter of solution.

Overview

The basis for the COD test is that nearly all organic compounds can be fully oxidized to carbon dioxide with a strong oxidizing agent under acidic conditions. The amount of oxygen required to oxidize an organic compound to carbon dioxide, ammonia, and water is given by:

$$C_nH_aO_bN_c + \left(n + \frac{a}{4} - \frac{b}{2} - \frac{3}{4}c\right)O_2 \rightarrow nCO_2 + \left(\frac{a}{2} - \frac{3}{2}c\right)H_2O + cNH_3$$

This expression does not include the oxygen demand caused by nitrification, the oxidation of ammonia into nitrate:

$$NH_3 + 2O_2 \rightarrow NO_3^- + H_3O^+$$

Dichromate, the oxidizing agent for COD determination, does not oxidize ammonia into nitrate, so nitrification is not included in the standard COD test.

The International Organization for Standardization describes a standard method for measuring chemical oxygen demand in ISO 6060 .

Using Potassium Dichromate

Potassium dichromate is a strong oxidizing agent under acidic conditions. Acidity is usually achieved by the addition of sulfuric acid. The reaction of potassium dichromate with organic compounds is given by:

$$C_nH_aO_bN_c + dCr_2O_7^{2-} + (8d + c)H^+ \rightarrow nCO_2 + \frac{a+8d-3c}{2}H_2O + cNH_4{+} + 2dCr^{3+}$$

where $d = 2n/3 + a/6 - b/3 - c/2$. Most commonly, a 0.25 N solution of potassium dichromate is used for COD determination, although for samples with COD below 50 mg/L, a lower concentration of potassium dichromate is preferred.

In the process of oxidizing the organic substances found in the water sample, potassium dichromate is reduced (since in all redox reactions, one reagent is oxidized and the other is reduced), forming Cr^{3+}. The amount of Cr^{3+} is determined after oxidization is complete, and is used as an indirect measure of the organic contents of the water sample.

Measurement of Excess

For all organic matter to be completely oxidized, an excess amount of potassium dichromate (or any oxidizing agent) must be present. Once oxidation is complete, the amount of excess potassium dichromate must be measured to ensure that the amount of Cr^{3+} can be determined with accuracy. To do so, the excess potassium dichromate is titrated with ferrous ammonium sulfate (FAS) until all of the excess oxidizing agent has been reduced to Cr^{3+}. Typically, the oxidation-reduction indicator ferroin is added during this titration step as well. Once all the excess dichromate has been reduced, the ferroin indicator changes from blue-green to a reddish brown. The amount of ferrous ammonium sulfate added is equivalent to the amount of excess potassium dichromate added to the original sample. Note: Ferroin indicator is bright red from commercially prepared sources, but when added to a digested sample containing potassium dichromate it exhibits a green hue. During the titration the color of the indicator changes from a green hue to a bright blue hue to a reddish brown upon reaching the endpoint. Ferroin indicator changes from red to pale blue when oxidized.

Preparation of Ferroin Indicator Reagent

A solution of 1.485 g 1,10-phenanthroline monohydrate is added to a solution of 695 mg $FeSO_4{\cdot}7H_2O$ in distilled water, and the resulting red solution is diluted to 100 mL.

Calculations

The following formula is used to calculate COD:

$$COD = \frac{8000(b - s)n}{\text{sample volume}}$$

where b is the volume of FAS used in the blank sample, s is the volume of FAS in the original sample, and n is the normality of FAS. If milliliters are used consistently for volume measurements, the result of the COD calculation is given in mg/L.

The COD can also be estimated from the concentration of oxidizable compound in the sample, based on its stoichiometric reaction with oxygen to yield CO_2 (assume all C goes to CO_2), H_2O (assume all H goes to H_2O), and NH_3 (assume all N goes to NH_3), using the following formula:

$$COD = (C/FW)\cdot(RMO)\cdot 32$$

Where

C = Concentration of oxidizable compound in the sample,

FW = Formula weight of the oxidizable compound in the sample,

RMO = Ratio of the # of moles of oxygen to # of moles of oxidizable compound in their reaction to CO_2, water, and ammonia

For example, if a sample has 500 wppm of phenol:

$$C_6H_5OH + 7O_2 \rightarrow 6CO_2 + 3H_2O$$

$$COD = (500/94)\cdot 7\cdot 32 = 1191 \text{ wppm}$$

Inorganic Interference

Some samples of water contain high levels of oxidizable inorganic materials which may interfere with the determination of COD. Because of its high concentration in most wastewater, chloride is often the most serious source of interference. Its reaction with potassium dichromate follows the equation:

$$6Cl^- + Cr_2O_7^{2-} + 14H^+ \rightarrow 3Cl_2 + 2Cr^{3+} + 7H_2O$$

Prior to the addition of other reagents, mercuric sulfate can be added to the sample to eliminate chloride interference.

The following table lists a number of other inorganic substances that may cause interference. The table also lists chemicals that may be used to eliminate such interference, and the compounds formed when the inorganic molecule is eliminated.

Inorganic molecule	Eliminated by	Elimination forms
Chloride	Mercuric sulfate	Mercuric chloride complex
Nitrite	Sulfamic acid	N_2 gas
Ferrous iron	-	-
Sulfides	-	-

Government Regulation

Many governments impose strict regulations regarding the maximum chemical oxygen demand allowed in waste water before they can be returned to the environment. For example, in Switzerland, a maximum oxygen demand between 200 and 1000 mg/L must be reached before waste water or industrial water can be returned to the environment .

History

For many years, the strong oxidizing agent potassium permanganate ($KMnO_4$) was used for measuring chemical oxygen demand. Measurements were called *oxygen consumed* from permanganate, rather than the *oxygen demand* of organic substances. Potassium permanganate's effectiveness at oxidizing organic compounds varied widely, and in many cases biochemical oxygen demand (BOD) measurements were often much greater than results from COD measurements. This indicated that potassium permanganate was not able to effectively oxidize all organic compounds in water, rendering it a relatively poor oxidizing agent for determining COD.

Since then, other oxidizing agents such as ceric sulphate, potassium iodate, and potassium dichromate have been used to determine COD. Of these, potassium dichromate ($K_2Cr_2O_7$) has been shown to be the most effective: it is relatively cheap, easy to purify, and is able to nearly completely oxidize almost all organic compounds.

In these methods, a fixed volume with a known excess amount of the oxidant is added to a sample of the solution being analyzed. After a refluxing digestion step, the initial concentration of organic substances in the sample is calculated from a titrimetric or spectrophotometric determination of the oxidant still remaining in the sample. As with all colorimetric methods blanks are used to control for contamination by outside material.

Biochemical Oxygen Demand

Biochemical oxygen demand (BOD, also called biological oxygen demand) is the amount of dissolved oxygen needed (i.e., demanded) by aerobic biological organisms to break down organic material present in a given water sample at certain temperature over a specific time period. The BOD value is most commonly expressed in milligrams of oxygen consumed per litre of sample during 5 days of incubation at 20 °C and is often used as a surrogate of the degree of organic pollution of water.

BOD test bottles at the laboratory of a wastewater treatment plant.

BOD can be used as a gauge of the effectiveness of wastewater treatment plants. It is listed as a conventional pollutant in the U.S. Clean Water Act.

BOD is similar in function to chemical oxygen demand (COD), in that both measure the amount of organic compounds in water. However, COD is less specific, since it measures everything that can be chemically oxidized, rather than just levels of biodegradable organic matter.

Background

Most natural waters contain small quantities of organic compounds. Aquatic microorganisms have evolved to use some of these compounds as food. Microorganisms living in oxygenated waters use dissolved oxygen to oxidatively degrade the organic compounds, releasing energy which is used for growth and reproduction. Populations of these microorganisms tend to increase in proportion to the amount of food available. This microbial metabolism creates an oxygen demand proportional to the amount of organic compounds useful as food. Under some circumstances, microbial metabolism can consume dissolved oxygen faster than atmospheric oxygen can dissolve into the water or the autotrophic community (algae, cyanobacteria and macrophytes) can produce. Fish and aquatic insects may die when oxygen is depleted by microbial metabolism.

Biochemical oxygen demand is the amount of oxygen required for microbial metabolism of organic compounds in water. This demand occurs over some variable period of time depending on temperature, nutrient concentrations, and the enzymes available to indigenous microbial populations. The amount of oxygen required to completely oxidize the organic compounds to carbon dioxide and water through generations of microbial growth, death, decay, and cannibalism is total biochemical oxygen demand (total BOD). Total BOD is of more significance to food webs than to water quality. Dissolved oxygen depletion is most likely to become evident during the initial aquatic microbial population explosion in response to a large amount of organic material. If the microbial population deoxygenates the water, however, that lack of oxygen imposes a limit on population growth of aerobic aquatic microbial organisms resulting in a longer term food surplus and oxygen deficit.

A standard temperature at which BOD testing should be carried out was first proposed by the Royal Commission on Sewage Disposal in its eighth report in 1912:

"(c) An effluent in order to comply with the general standard must not contain as discharged more than 3 parts per 100,000 of suspended matter, and with its suspended matters included must not take up at 65°F more than 2.0 parts per 100,000 of dissolved oxygen in 5 days. This general standard should be prescribed either by Statute or by order of the Central Authority, and should be subject to modifications by that Authority after an interval of not less than ten years.

This was later standardised at 68 °F and then 20 °C. This temperature may be significantly different from the temperature of the natural environment of the water being tested.

Although the Royal Commission on Sewage Disposal proposed 5 days as an adequate test period for rivers of the United Kingdom of Great Britain and Ireland, longer periods were investigated for North American rivers. Incubation periods of 1, 2, 5, 10 and 20 days were being used into the mid-20th century. Keeping dissolved oxygen available at their chosen temperature, investigators found up to 99 percent of total BOD was exerted within 20 days, 90 percent within 10 days, and approximately 68 percent within 5 days. Variable microbial population shifts to nitrifying bacteria limit test reproducibility for periods greater than 5 days. The 5-day test protocol with acceptably reproducible results emphasizing carbonaceous BOD has been endorsed by the United States Environmental Protection Agency. This 5-day BOD test result may be described as the amount of oxygen required for aquatic microorganisms to stabilize decomposable organic matter under aerobic conditions. Stabilization, in this context, may be perceived in general terms as the conversion of food to living aquatic fauna. Although these fauna will continue to exert biochemical oxygen demand as they die, that tends to occur within a more stable evolved ecosystem including higher trophic levels.

Taking samples from the influent raw wastewater stream for BOD measurements at a wastewater treatment plant in Haran-Al-Awamied near Damascus in Syria

Typical Values

Most pristine rivers will have a 5-day carbonaceous BOD below 1 mg/L. Moderately polluted rivers may have a BOD value in the range of 2 to 8 mg/L. Rivers may be considered severely polluted when BOD values exceed 8 mg/L. Municipal sewage that is efficiently treated by a three-stage process would have a value of about 20 mg/L or less. Untreated sewage varies, but averages around 600 mg/L in Europe and as low as 200 mg/L in the U.S., or where there is severe groundwater or surface water Infiltration/Inflow. The generally lower values in the U.S. derive from the much greater water use per capita than in other parts of the world.

Methods

There are two commonly recognized methods for the measurement of BOD.

Dilution Method

This standard method is recognized by U.S. EPA, which is labeled Method 5210B in the Standard Methods for the Examination of Water and Wastewater In order to obtain BOD_5, dissolved oxygen (DO) concentrations in a sample must be measured before and after the incubation period, and appropriately adjusted by the sample corresponding dilution factor. This analysis is performed using 300 ml incubation bottles in which buffered dilution water is dosed with seed microorganisms and stored for 5 days in the dark room at 20 °C to prevent DO production via photosynthesis. In addition to the various dilutions of BOD samples, this procedure requires dilution water blanks, glucose glutamic acid (GGA) controls, and seed controls. The dilution water blank is used to confirm the quality of the dilution water that is used to dilute the other samples. This is necessary because impurities in the dilution water may cause significant alterations in the results. The GGA control is a standardized solution to determine the quality of the seed, where its recommended BOD_5 concentration is 198 mg/l ± 30.5 mg/l. For measurement of carbonaceous BOD (cBOD), a nitrification inhibitor is added after the dilution water has been added to the sample. The inhibitor hinders the oxidation of ammonia nitrogen, which supplies the nitrogenous BOD (nBOD). When performing the BOD_5 test, it is conventional practice to measure only cBOD because nitrogenous demand does not reflect the oxygen demand from organic matter. This is because nBOD is generated by the breakdown of proteins, whereas cBOD is produced by the breakdown of organic molecules.

BOD_5 is calculated by:

- Unseeded: $BOD_5 = \frac{(D_0 - D_5)}{P}$
- Seeded: $BOD_5 = \frac{(D_0 - D_5) - (B_0 - B_5)f}{P}$

where:

D_0 is the dissolved oxygen (DO) of the diluted solution after preparation (mg/l)

D_5 is the DO of the diluted solution after 5 day incubation (mg/l)

P is the decimal dilution factor

B_0 is the DO of diluted seed sample after preparation (mg/l)

B_5 is the DO of diluted seed sample after 5 day incubation (mg/l)

f is the ratio of seed volume in dilution solution to seed volume in BOD test on seed

Manometric Method

This method is limited to the measurement of the oxygen consumption due only to carbonaceous oxidation. Ammonia oxidation is inhibited.

The sample is kept in a sealed container fitted with a pressure sensor. A substance that absorbs carbon dioxide (typically lithium hydroxide) is added in the container above the sample level. The sample is stored in conditions identical to the dilution method. Oxygen is consumed and, as ammonia oxidation is inhibited, carbon dioxide is released. The total amount of gas, and thus the pressure, decreases because carbon dioxide is absorbed. From the drop of pressure, the sensor electronics computes and displays the consumed quantity of oxygen.

The main advantages of this method compared to the dilution method are:

- simplicity: no dilution of sample required, no seeding, no blank sample.
- direct reading of BOD value.
- continuous display of BOD value at the current incubation time.

Alternative Methods

Biosensor

An alternative to measure BOD is the development of biosensors, which are devices for the detection of an analyte that combines a biological component with a physicochemical detector component. Enzymes are the most widely used biological sensing elements in the fabrication of biosensors. Their application in biosensor construction is limited by the tedious, time consuming and costly enzyme purification methods. MIcroorganisms provide an ideal alternative to these bottlenecks.

The vast variety of micro organisms are relatively easy to maintain in pure cultures, grow and harvest at low cost. Moreover, the use of microbes in biosensor field has opened up new possibilities and advantages such as ease of handling, preparation and low cost of device. A number of pure cultures, e.g. *Trichosporon cutaneum, Bacillus cereus, Klebsiella oxytoca, Pseudomonas sp.* etc. individually, have been used by many workers for the construction of BOD biosensor. On the other hand, many workers have immobilized activated sludge, or a mixture of two or three bacterial species and on various membranes for the construction of BOD biosensor. The most commonly used membranes were polyvinyl alcohol, porous hydrophilic membranes etc.

A defined microbial consortium can be formed by conducting a systematic study, i.e. pre-testing of selected micro-organisms for use as a seeding material in BOD analysis of a wide variety of industrial effluents. Such a formulated consortium can be immobilized on suitable membrane, i.e. charged nylon membrane useful for BOD estimation. Suitability of charges nylon membrane lies in the specific binding between negatively charged bacterial cell and positively charged nylon membrane. So the advantages of the nylon membrane over the other membranes are : The dual binding, i.e. Adsorption as well as entrapment, thus resulting in a more stable immobilized membrane. Such specific Microbial consortium based BOD analytical devices, may find great application in monitoring of the degree of pollutional strength, in a wide variety of Industrial waste water within a very short time.

Biosensors can be used to indirectly measure BOD via a fast (usually <30 min) to be determined BOD substitute and a corresponding calibration curve method (pioneered by Karube et al., 1977). Consequently, biosensors are now commercially available, but they do have several limitations such as their high maintenance costs, limited run lengths due to the need for reactivation, and the inability to respond to changing quality characteristics as would normally occur in wastewater treatment streams; e.g. diffusion processes of the biodegradable organic matter into the membrane and different responses by different microbial species which lead to problems with the reproducibility of result (Praet et al., 1995). Another important limitation is the uncertainty associated with the calibration function for translating the BOD substitute into the real BOD (Rustum *et al.*, 2008).

Fluorescent RedOx Indicator

A surrogate to BOD_5 has been developed using a resazurin derivative which reveals the extent of oxygen uptake by micro-organisms for organic matter mineralization. A cross-validation performed on 109 samples in Europe and the United-States showed a strict statistical equivalence between results from both methods. The French start-up Envolure (Montpellier, France) offers the kit ENVERDI which enables the users to perform up to 40 BOD_5 simultaneously in 48 hours in a single 96-wells microplate.

Software Sensor

Rustum et al. (2008) proposed the use of the KSOM to develop intelligent models for making rapid inferences about BOD using other easy to measure water quality parameters, which, unlike BOD, can be obtained directly and reliably using on-line hardware sensors. This will make the use of BOD for on-line process monitoring and control a more plausible proposition. In comparison to other data-driven modeling paradigms such as multi-layer perceptrons artificial neural networks (MLP ANN) and classical multi-variate regression analysis, the KSOM is not negatively affected by missing data. Moreover, time sequencing of data is not a problem when compared to classical time series analysis.

Dissolved Oxygen Probes: Membrane and Luminescence

Since the publication of a simple, accurate and direct dissolved oxygen analytical procedure by Winkler, the analysis of dissolved oxygen levels for water has been key to the determination of surface water purity and ecological wellness. The Winkler method is still one of only two analytical techniques used to calibrate oxygen electrode meters; the other procedure is based on oxygen solubility at saturation as per Henry's law. Though many researchers have refined the Winkler analysis to dissolved oxygen levels in the low PPB range, the method does not lend itself to automation.

The development of an analytical instrument that utilizes the reduction-oxidation (redox) chemistry of oxygen in the presence of dissimilar metal electrodes was introduced during the 1950s. This redox electrode utilized an oxygen-permeable membrane to allow the diffusion of the gas into an electrochemical cell and its concentration determined by polarographic or galvanic electrodes. This analytical method is sensitive and accurate to down to levels of ± 0.1 mg/l dissolved oxygen. Calibration of the redox electrode of this membrane electrode still requires the use of the Henry's law table or the Winkler test for dissolved oxygen.

During the last two decades, a new form of electrode was developed based on the luminescence emission of a photo active chemical compound and the quenching of that emission by oxygen. This quenching photophysics mechanism is described by the Stern–Volmer equation for dissolved oxygen in a solution:

$$I_0 / I = 1 + K_{SV} [O_2]$$

- I : Luminescence in the presence of oxygen
- I_0 : Luminescence in the absence of oxygen
- K_{SV} : Stern-Volmer constant for oxygen quenching
- $[O_2]$: Dissolved oxygen concentration

The determination of oxygen concentration by luminescence quenching has a linear response over a broad range of oxygen concentrations and has excellent accuracy and reproducibility. There are several recognized EPA methods for the measurement of dissolved oxygen for BOD, including the following methods:

- Standard Methods for the Examination of Water and Wastewater, Method 4500 O
- In-Situ Inc. Method 1003-8-2009 Biochemical Oxygen Demand (BOD) Measurement by Optical Probe.

Test Limitations

The test method involves variables limiting reproducibility. Tests normally show observations varying plus or minus ten to twenty percent around the mean.

Toxicity

Some wastes contain chemicals capable of suppressing microbiological growth or activity. Potential sources include industrial wastes, antibiotics in pharmaceutical or medical wastes, sanitizers in food processing or commercial cleaning facilities, chlorination disinfection used following conventional sewage treatment, and odor-control formulations used in sanitary waste holding tanks in passenger vehicles or portable toilets. Suppression of the microbial community oxidizing the waste will lower the test result.

Appropriate Microbial Population

The test relies upon a microbial ecosystem with enzymes capable of oxidizing the available organic material. Some waste waters, such as those from biological secondary sewage treatment, will already contain a large population of microorganisms acclimated to the water being tested. An appreciable portion of the waste may be utilized during the holding period prior to commencement of the test procedure. On the other hand, organic wastes from industrial sources may require specialized enzymes. Microbial populations from standard seed sources may take some time to produce those enzymes. A specialized seed culture may be appropriate to reflect conditions of an evolved ecosystem in the receiving waters.

History

The *Royal Commission on River Pollution*, which was established in 1865 and the formation of the *Royal Commission on Sewage Disposal* in 1898 led to the selection in 1908 of BOD_5 as the definitive test for organic pollution of rivers. Five days was chosen as an appropriate test period because this is supposedly the longest time that river water takes to travel from source to estuary in the U.K.. In its sixth report the Royal Commission recommended that the standard set should be 15 parts by weight per million of water. However, in the Ninth report the commission had revised the recommended standard :

"An effluent taking up 2–0 parts dissolved oxygen per 100,000 would be found by a simple calculation to require dilution with at least 8 volumes of river water taking up 0.2 part if the resulting mixture was not to take up more than 0.4 part. Our experience indicated that in a large majority of cases the volume of river water would exceed 8 times the volume of effluent, and that the figure of 2–0 parts dissolved oxygen per 100,000, which had been shown to be practicable, would be a safe figure to adopt for the purposes of a general standard, taken in conjunction with the condition that the effluent should not contain more than 3–0 parts per 100,000 of suspended solids."

This was the cornerstone 20:30 (BOD:Suspended Solids) + full nitrification standard which was used as a yardstick in the U.K. up to the 1970s for sewage works effluent quality.

The United States includes BOD effluent limitations in its secondary treatment regulations. Secondary sewage treatment is generally expected to remove 85 percent of the BOD measured in sewage and produce effluent BOD concentrations with a 30-day average of less than 30 mg/L and a 7-day average of less than 45 mg/L. The regulations also describe "treatment equivalent to secondary treatment" as removing 65 percent of the BOD and producing effluent BOD concentrations with a 30-day average less than 45 mg/L and a 7-day average less than 65 mg/L.

Total Dissolved Solids

Bottled mineral water usually contains higher TDS levels than tap water

Total dissolved solids (TDS) is a measure of the combined content of all inorganic and organic substances contained in a liquid in molecular, ionized or micro-granular (colloidal sol) suspended form. Generally the operational definition is that the solids must be small enough to survive filtration through a filter with two-micrometer (nominal size, or smaller) pores. Total dissolved solids are normally discussed only for freshwater systems, as salinity includes some of the ions constituting the definition of TDS. The principal application of TDS is in the study of water quality for streams, rivers and lakes, although TDS is not generally considered a primary pollutant (e.g. it is not deemed to be associated with health effects) it is used as an indication of aesthetic characteristics of drinking water and as an aggregate indicator of the presence of a broad array of chemical contaminants.

Primary sources for TDS in receiving waters are agricultural and residential runoff, clay rich mountain waters, leaching of soil contamination and point source water pollution discharge from industrial or sewage treatment plants. The most common chemical constituents are calcium, phosphates, nitrates, sodium, potassium and chloride, which are found in nutrient runoff, general stormwater runoff and runoff from snowy climates where road de-icing salts are applied. The chemicals may be cations, anions, molecules or agglomerations on the order of one thousand or fewer molecules, so long as a soluble micro-granule is formed. More exotic and harmful elements of TDS are pesticides arising from surface runoff. Certain naturally occurring total dissolved solids arise from the weathering and dissolution of rocks and soils. The United States has established a secondary water quality standard of 500 mg/l to provide for palatability of drinking water.

Total dissolved solids are differentiated from total suspended solids (**TSS**), in that the latter cannot pass through a sieve of two micrometers and yet are indefinitely suspended in solution. The term "settleable solids" refers to material of any size that will not remain suspended or dissolved in a holding tank not subject to motion, and excludes both TDS and TSS. Settleable solids may include larger particulate matter or insoluble molecules.

Measurement

The two principal methods of measuring total dissolved solids are gravimetric analysis and conductivity. Gravimetric methods are the most accurate and involve evaporating the liquid solvent and measuring the mass of residues left. This method is generally the best, although it is time-consuming. If inorganic salts comprise the great majority of TDS, gravimetric methods are appropriate.

Electrical conductivity of water is directly related to the concentration of dissolved ionized solids in the water. Ions from the dissolved solids in water create the ability for that water to conduct an electric current, which can be measured using a conventional conductivity meter or TDS meter. When correlated with laboratory TDS measurements, conductivity provides an approximate value for the TDS concentration, usually to within ten-percent accuracy.

The relationship of TDS and specific conductance of groundwater can be approximated by the following equation:

$$TDS = k_e EC$$

where TDS is expressed in mg/L and EC is the electrical conductivity in microsiemens per centimeter at 25 °C. The correlation factor k_e varies between 0.55 and 0.8.

Hydrological Simulation

Pyramid Lake, Nevada receives dissolved solids from the Truckee River.

Hydrologic transport models are used to mathematically analyze movement of TDS within river systems. The most common models address surface runoff, allowing variation in land use type, topography, soil type, vegetative cover, precipitation, and land management practice (e.g. the application rate of a fertilizer). Runoff models have evolved to a good degree of accuracy and permit the evaluation of alternative land management practices upon impacts to stream water quality.

Basin models are used to more comprehensively evaluate total dissolved solids within a catchment basin and dynamically along various stream reaches. The DSSAM model was developed by the U.S. Environmental Protection Agency (EPA). This hydrology transport model is actually based upon the pollutant-loading metric called "Total Maximum Daily Load" (TMDL), which addresses TDS and other specific chemical pollutants. The success of this model contributed to the Agency's broadened commitment to the use of the underlying TMDL protocol in its national policy for management of many river systems in the United States.

Practical Implications

High levels of total dissolved solids do not correlate to hard water, as water softeners do not reduce TDS. Water softners remove magnesium and calcium ions, which cause hard water, but these ions are replaced with an equal number of sodium or potassium ions. This leaves overall TDS unchanged. Hard water can cause scale buildup in pipes, valves, and filters, reducing performance and adding to system maintenance costs. These effects can be seen in aquariums, spas, swimming pools, and reverse osmosis water treatment systems. Typically, in these applications, total dissolved solids are tested frequently, and filtration membranes are checked in order to prevent adverse effects.

Aquarium at Bristol Zoo, England. Maintenance of filters becomes costly with high TDS.

In the case of hydroponics and aquaculture, TDS is often monitored in order to create a water quality environment favorable for organism productivity. For freshwater oysters, trouts, and other high value seafood, highest productivity and economic returns are achieved by mimicking the TDS and pH levels of each species' native environment. For hydroponic uses, total dissolved solids is considered one of the best indices of nutrient availability for the aquatic plants being grown.

Because the threshold of acceptable aesthetic criteria for human drinking water is 500 mg/l, there is no general concern for odor, taste, and color at a level much lower than is required for harm. A number of studies have been conducted and indicate various species' reactions range from intolerance to outright toxicity due to elevated TDS. The numerical results must be interpreted cautiously, as true toxicity outcomes will relate to specific chemical constituents. Nevertheless, some numerical information is a useful guide to the nature of risks in exposing aquatic organisms or terrestrial animals to high TDS levels. Most aquatic ecosystems involving mixed fish fauna can tolerate TDS levels of 1000 mg/l.

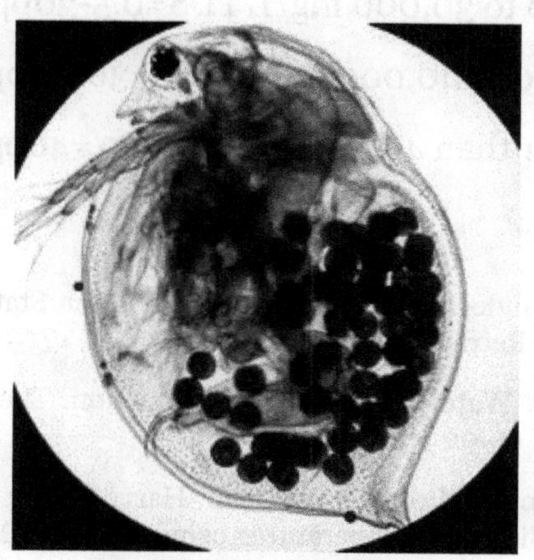

Daphnia magna with eggs

The Fathead minnow (*Pimephales promelas*), for example, realizes an LD_{50} concentration of 5600 ppm based upon a 96-hour exposure. LD50 is the concentration required to produce a lethal effect on 50 percent of the exposed population. *Daphnia magna*, a good example of a primary member of the food chain, is a small planktonic crustacean, about 0.5 mm in length, having an LD50 of about 10,000 ppm TDS for a 96 hour exposure.

Spawning fishes and juveniles appear to be more sensitive to high TDS levels. For example, it was found that concentrations of 350 mg/l TDS reduced spawning of Striped bass (*Morone saxatilis*) in the San Francisco Bay-Delta region, and that concentrations below 200 mg/l promoted even healthier spawning conditions. In the Truckee River, EPA found that juvenile Lahontan cutthroat trout were subject to higher mortality when exposed to thermal pollution stress combined with high total dissolved solids concentrations.

For terrestrial animals, poultry typically possess a safe upper limit of TDS exposure of approximately 2900 mg/l, whereas dairy cattle are measured to have a safe upper limit of about 7100 mg/l. Research has shown that exposure to TDS is compounded in toxicity when other stressors are present, such as abnormal pH, high turbidity, or reduced dissolved oxygen with the latter stressor acting only in the case of animalia.

In countries with often-unsafe/unclean tap water supplies, such as much of India, the TDS of drinking water is often checked by technicians to gauge how effectively their RO/Water Filtration devices are working. While TDS readings will not give an answer as to the amount of microorganisms present in a sample of water, they can get a good idea as to the efficiency of the filter by how much TDS is present.

Water Classification

Water can be classified by the level of TDS in the water:

- Fresh water: less than 500 mg/L TDS=0.5ppt
- Brackish water: 500 to 30,000 mg/L TDS=0.5-30ppt
- Saline water: 30,000 to 40,000 mg/L TDS=30-40ppt
- Hypersaline: greater than 40,000 mg/L TDS>=40ppt

References

- W. Adam Sigler, Jim Bauder. "TDS Fact Sheet". Montana State University. Archived from the original on 2015-04-29. Retrieved 23 January 2015.
- Boyd, Claude E. (1999). Water Quality: An Introduction. The Netherlands: Kluwer Academic Publishers Group. ISBN 0-7923-7853-9.
- Hogan, C. Michael; Patmore, Leda C.; Seidman, Harry (August 1973). "Statistical Prediction of Dynamic Thermal Equilibrium Temperatures using Standard Meteorological Data Bases". EPA. Retrieved 2016-02-15.

5

Bioindicator: An Integrated Study

Bioindicators are biological species whose functions reveal the qualitative status of the environment. A biological monitor is an organism that provides quantitative information and this information provides qualitative information regarding the environment. This chapter strategically encompasses and incorporates the major components and the main concepts of bioindictors, providing a complete understanding.

Bioindicator

A bioindicator is any biological species (an "indicator species") or group of species whose function, population, or status can reveal the qualitative status of the environment. For example, copepods and other small water crustaceans that are present in many water bodies can be monitored for changes (biochemical, physiological, or behavioural) that may indicate a problem within their ecosystem. Bioindicators can tell us about the cumulative effects of different pollutants in the ecosystem and about how long a problem may have been present, which physical and chemical testing cannot.

Caddisfly (order Trichoptera), a macroinvertebrate used as an indicator of water quality.

A biological monitor, or biomonitor, can be defined as an organism that provides quantitative information on the quality of the environment around it. Therefore, a good biomonitor will indicate the presence of the pollutant and also attempt to provide additional information about the amount and intensity of the exposure.

Overview

A bio indicator is an organism or biological response that reveals the presence of the pollutants by the occurrence of typical symptoms or measurable responses, and is therefore more qualitative. These organisms (or communities of organisms) deliver information on alterations in the environment or the quantity of environmental pollutants by changing in one of the following ways: physiologically, chemically or behaviourally. The information can be deduced through the study of:

1. their content of certain elements or compounds
2. their morphological or cellular structure
3. metabolic-biochemical processes
4. behaviour, or
5. population structure(s).

The importance and relevance of biomonitors, rather than man-made equipment, is justified by the statement: "There is no better indicator of the status of a species or a system than a species or system itself."

The use of a biomonitor is described as biological monitoring (*abbr.* biomonitoring) and is the use of the properties of an organism to obtain information on certain aspects of the biosphere. Biomonitoring of air pollutants can be passive or active. Passive methods observe plants growing naturally within the area of interest. Active methods detect the presence of air pollutants by placing test plants of known response and genotype into the study area.

Bioaccumulative indicators are frequently regarded as biomonitors.

Depending on the organism selected and their use, there are several types of bio-indicators.

Plant Indicators

The lichen *Lobaria pulmonaria* is sensitive to air pollution.

The presence or absence of certain plant or other vegetative life in an ecosystem can provide important clues about the health of the environment: environmental preservation.

There are several types of plant and fungi biomonitors, including mosses, lichens, tree bark, bark pockets, tree rings, leaves, and fungi.

- Lichens are organisms comprising both fungi and algae. They are found on rocks and tree trunks, and they respond to environmental changes in forests, including changes in forest structure – conservation biology, air quality, and climate. The disappearance of lichens in a forest may indicate environmental stresses, such as high levels of sulfur dioxide, sulfur-based pollutants, and nitrogen oxides.
- The composition and total biomass of algal species in aquatic systems serves as an important metric for organic water pollution and nutrient loading such as nitrogen and phosphorus.

There are genetically engineered organisms, that can respond to toxicity levels in the environment; *e.g.*, a type of genetically engineered grass that grows a different colour if there are toxins in the soil.

Animal indicators and Toxins

An increase or decrease in an animal population may indicate damage to the ecosystem caused by pollution. For example, if pollution causes the depletion of important food sources, animal species dependent upon these food sources will also be reduced in number: population decline. Overpopulation can be the result of opportunistic species growth. In addition to monitoring the size and number of certain species, other mechanisms of animal indication include monitoring the concentration of toxins in animal tissues, or monitoring the rate at which deformities arise in animal populations, or their behaviour either directly in the field or in a lab.

Microbial Indicators and Chemical Pollutants

Microorganisms can be used as indicators of aquatic or terrestrial ecosystem health. Found in large quantities, microorganisms are easier to sample than other organisms. Some microorganisms will produce new proteins, called stress proteins, when exposed to contaminants such as cadmium and benzene. These stress proteins can be used as an early warning system to detect changes in levels of pollution.

Microbial Indicators in Oil and Gas Exploration

Microbial Prospecting for oil and gas (MPOG) is often used to identify prospective areas for oil and gas occurrences. In many cases oil and gas is known to seep toward the surface as a hydrocarbon reservoir will usually leak or have leaked towards the surface through buoyancy forces overcoming sealing pressures. These hydrocarbons can alter the chemical and microbial occurrences found in the near surface soils or can be picked up directly. Techniques used for MPOG include DNA analysis, simple bug counts after culturing a soil sample in a hydrocarbon based medium or by looking at the consumption of hydrocarbon gases in a culture cell.

Microalgae as Bio-indicators for Water Quality

Microalgae have gained attention in the recent years due to several reasons because of their greater sensitivity to pollutants than many other organisms. In addition they occur abundantly in nature, they are an essential component in very many food webs, they are easy to culture and to use in assays and there are few if any ethical issues involved in their use.

Euglena gracilis is a motile freshwater photosynthetic flagellate. Although *Euglena* is rather tolerant to acidity, it responds rapidly and sensitively to environmental stresses such as heavy metals or inorganic and organic compounds. Typical responses are the inhibition of movement and the change of orientation parameters. Moreover, this organism is very easy to handle and grows, making it a very useful tool for eco-toxicological assessments. One very useful particularity of this organism is the gravitactic orientation, which is very sensitive to pollutants.

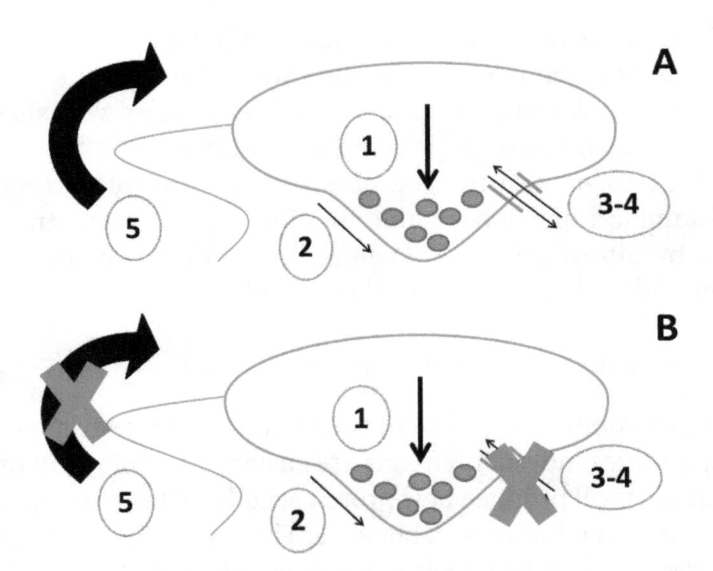

Gravitactic mechanism of the microalgae Euglena gracilis (A) in the absence and (B) in the presence of pollutants.

The gravireceptors are impaired by pollutants such as heavy metals and organic or inorganic compounds. Therefore, the presence of such substances is associated with random movement of the cells in the water column. For short term tests, gravitactic orientation of *E. gracilis* is very sensitive. Other species such as *Paramecium biaurelia* also use gravitactic orientation.

ECOTOX

ECOTOX is an automatic bioassay device used to test the quality of water samples, by the detection of toxic chemicals. It is small piece of hardware containing a miniaturized microscope linked to a camera, an observation cuvette, pumps to mix the water samples with the microalgae; everything being connected to a computer equipped with software. One of the biggest advantages of this device is the automated measurements and analysis, which reduces the risks of personal error. Moreover, it is easy to use, quite cheap and fast: only 10 min are necessary to test a water sample and the corresponding control. Examples of use are the test of seepage water or the determination of the efficiency of purification systems by testing treated waste water before and after purification. The determination of the samples quality is derived from analysis of several parameters related to the movement of the microalgae. All measurements are made automatically with real time image analysis. First the orientation behaviour of the cells is determined using two parameters: the percentage of cells moving upwards giving the direction of the movement and the r-value indicating the precision of the gravitactic orientation which varies from a random movement (r-value=0) to a single direction (r-value=1). Other important parameters are the velocity, the cell motility which represents the percentage of cells moving faster than the minimum velocity and the cell compactness giving information about the shape of the cell. All parameters are compared with a control sample of unpolluted tap water and the percentage of inhibition is calculated. An inhibition indicates the presence of a pollutant. Depending on the aim of the study, the EC50 (the concentration of sample which affects 50 percent of organisms) and the G-value (lowest dilution factor at which no-significant toxic affect can be measured), are calculated. From all those parameters, the gravitactic orientation represented with upward swimming and r-value is the most sensitive.

Macroinvertebrate Bio-indicators

Macroinvertebrates are useful and convenient indicators of the ecological health of a water body. They are almost always present, and are easy to sample and identify. The sensitivity of the range of macroinvertebrates found will enable an objective judgement of the ecological condition to be made. Tolerance values are commonly used to assess water pollution.

In Australia, the SIGNAL method has been developed and is used by researchers and community "Waterwatch" groups to monitor water health.

In Europe, a remote online biomonitoring system was designed in 2006. It is based on bivalve molluscs and the exchange of real time data between a remote intelligent device in the field (able to work for more than 1 year without *in-situ* human intervention) and a data centre designed to capture, process and distribute on the web information derived from the data. The technique relates bivalve behaviour, specifically shell gaping activity, to water quality changes. This technology has been successfully used for the assessment of coastal water quality in various countries (France, Spain, Norway, Russia, Svalbard (Ny Alesund) and New Caledonia).

In the United States, the Environmental Protection Agency (EPA) has published *Rapid Bioassessment Protocols,* based on macroinvertebrates, as well as periphyton and fish. These protocols are used by many federal, state and local government agencies to design biosurveys for assessment of water quality. Volunteer stream monitoring organizations around the U.S., working in cooperation with government agencies, typically use macroinvertebrate methods. The species identification procedures are conducted in the field without the use of specialized equipment, and the techniques can be easily taught in volunteer training sessions.

In South Africa, the Southern African Scoring System (SASS) method was developed as a rapid bioassessment technique, based on benthic macroinvertebrates, and is used for the assessment of water quality in Southern African rivers. The SASS aquatic biomonitoring tool has been refined over the past 30 years and is now on the fifth version (SASS5) which has been specifically modified in accordance with international standards, namely the ISO/IEC 17025 protocol. The SASS5 method is used by the South African Department of Water Affairs as a standard method for River Health Assessment, which feeds the national River Health Programme and the national Rivers Database.

The imposex phenomenon in the dog conch species of sea snail leads to the abnormal development of a penis in females, but does not cause sterility. Because of this, the species has been suggested as a good indicator of pollution with organic man-made tin compounds in Malaysian ports.

Indicator Plant

Since a plant species or plant community acts as a measure of environmental conditions, it is referred to as biological indicators, bioindicators or phytoindicators. Plants which indicate some specific conditions of environment are called plant indicators.

Plants

Potential Productivity of Land

Forest serve as good indicators of land productivity. For example- vegetative growth of trees like *Quercus. marilandica, Q. stellata* are comparatively poor on lowland or sterile sandy soil than the normal soil in which they grow under conditions.

Climate

Evergreen forests indicate high rainfall in winter as well as summer; Sclerphyllous vegetation indicate heavy rainfall in winter and low during summer; Grassland indicate heavy rains during summer and low during winter; Xerophytic vegetation indicate a very low or no rainfall in the year.

Soil Characteristics

Luxuriant (abundant) growth of some taller and deeply rooted legumes like *Psoralea* indicates a sandy loam type of soil, whereas the presence of grasses *Andropogon* indicates sandy soil. *Rumex acetosella* indicates an acid grassland soil, whereas *Spermacoce stricta* the iron rich soil in the area. *Shorea robusta, Cassia obtusifolia, Geranium sp.* and *Impatiens sp.* indicate proper aeration of soil.

Fires

Some plants as *Agrostis, Eupilobium, Pium, Populus, Pteris, and Pyronema* dominate in areas destroyed by fires. Particular species of *Pteridium* indicate burnt and highly disturbed coniferous forests.

Pollution

Plants like *Utricularia, Chara, Wolffia* grow in polluted waters. Bacteria like *E. coli* also indicate water pollution. Presence of diatoms in water indicates pollution by sewage.

Lichen

i) Indicators of Radioactive Particles

Dead lichen thalii are capable of absorbing F, heavy metals including SO_2 and Pb. Lichens are also utilized for survey of long life nuclides (a distinct kind of atom or nucleus characterized by a specific number of protons and neutrons) like strontium (90_{Sr}) and cesium (137_{Cs}) released from nuclear explosions.

ii) Sensitivity to Air Pollution

Lichens can thus be used as reliable biological indicators of pollution. *Lecanora conizaeoides* is the most tolerant of all lichens to SO_2, thus occurs in city also. Thus lichens are used as pollution monitors.

Algae

1. *Ulva, Enteromorpha* are used in monitoring the water quality of estuaries (the tidal mouth of a large river).

2. Heavy metal pollution of water in monitored by *Cladophora, Stigeoclonium.*

3. *Chlorella* is used to monitor toxic substances in water.
4. *Duniella teritolecta, Cyclotella cryptic, Pavlova lutheri* are indicators of oil pollution.
5. Cyanobacteria like *Nostoc, Microscopium, Haplosiphon, Welwitschii* are the indicators of soil pesticides as dithane, deltan, BHC, aldrex, rogor.

Fungi and Bacteria

1. Fungi are used to monitor of oil pollution, *Scolecobasidium, Mortierella, Humicola, Verticillum* are able to utilize waste oil.
2. Bacteria like *Pseudomonas, Clostridium, and Streptococcus* are used in assessment (estimation) and prediction of changes in marine environment induced by human activities. *Pseudomonas* metabolizes oil and converts into harmless end product as CO_2 and H_2O.

Higher Plants

1. *Medicago sativa* grow in low SO_2.
2. *Rumex acetosella* grow in acidic soil.

Minerals

Many plants indicate the presence of characteristics minerals in the soils, theses plant are called metallocoles/metallophytes.

Gold- *Equisetum arvense*;

Diamond- *Vallozia candida*;

Silver-*Eriogonium ovalifolium*;

Much useful knowledge can be obtained about our land by observation of the wild plants and cover crops that are growing, and their condition. These include indicating fertility levels and potential nutrient deficiencies, waterlogging or compaction problems, pH levels and so on. When assessing land by observation of indicator species however it is good practice to observe plant communities or consistent populations of indicator species rather than individual specimens which may not be typical. Perennial weeds which may have colonised an area for some time are also a more reliable form of indicator than annual weeds which may have only been there for that year and thus indicate a temporary condition.

Viral Infection

Plant species will respond differentially to the presence of viruses due to variation in susceptibility. This can cause differences in observable symptoms as well as the magnitude of those symptoms. For instance, a plant may harbor a viral infection and show no observable symptoms yet still be negatively affected. Therefore, some plant species become useful as indicators because their symptoms will develop at a faster rate, are more obvious, or are more consistent for diagnosis. Common techniques of transferring a viral infection from a suspected plant to an indicator plant include grafting or sap transmission. However, ELISA testing, related serological tests, and direct electron microscope observation of viruses are more modern methods for detection.

Examples

- Fertile soil supports plants such as nettles, chickweed, groundsel and fat hen.
- Nitrogen-deficient conditions are indicated by the presence of nitrogen fixing legumes such as clovers or vetches.
- Bracken, plantains, sorrel, knapweeds, rhododendron and cranberries are amongst the plants that favour acidic conditions, while alkaline conditions tend to support populations of perennial sow thistle, bladder campion, and henbane.
- Waterlogged or poorly drained land is indicated by the presence of species such as mosses, creeping buttercup and horsetail, or bog loving plants including sedges, rushes, marsh marigold, or marsh orchid.

Environmental Indicator

Environmental indicators are simple measures that tell us what is happening in the environment. Since the environment is very complex, indicators provide a more practical and economical way to track the state of the environment than if we attempted to record every possible variable in the environment. For example, concentrations of ozone depleting substances (ODS) in the atmosphere, tracked over time, is a good indicator with respect to the environmental issue of stratospheric ozone depletion.

Environmental indicators have been defined in different ways but common themes exist.

"An environmental indicator is a numerical value that helps provide insight into the state of the environment or human health. Indicators are developed based on quantitative measurements or statistics of environmental condition that are tracked over time. Environmental indicators can be developed and used at a wide variety of geographic scales, from local to regional to national levels."

"A parameter or a value derived from parameters that describe the state of the environment and its impact on human beings, ecosystems and materials, the pressures on the environment, the driving forces and the responses steering that system. An indicator has gone through a selection and/or aggregation process to enable it to steer action."

Discussion

Environmental indicator criteria and frameworks have been used to help in their selection and presentation.

It can be considered, for example, that there are major subsets of environmental indicators in-line with the Pressure-State-Response model developed by the OECD. One subset of environmental indicators is the collection of ecological indicators which can include physical, biological and chemical measures such as atmospheric temperature, the concentration of ozone in the stratosphere or the number of breeding bird pairs in an area. These are also referred to as "state" indicators as their focus is on the state of the environment or conditions in the environment. A second subset is the collection of indicators that measure human activities or anthropogenic pressures, such as greenhouse gas emissions. These are also referred to as "pressure" indicators. Finally, there are indicators, such as the number of people serviced by sewage treatment, which track societal responses to environmental issues.

Environmental indicators, in turn, should be considered as a subset of sustainable development indicators which are meant to track the overall sustainability of a society with respect to its environmental, social and economic integrity and health.

A common framework spearheaded by the European Environment Agency is the "DPSIR" or "drivers, pressures, state, impact, response" framework. Drivers and pressures are indicators of the human activities and resulting pressures on the environment in the form of pollution or land-use change, for example. State and impact indicators are the resulting conditions in the environment and the implications for the health of ecosystems and humans. The response indicators measure the reaction of human society to the environmental issue. Criteria tend to focus on three key areas – scientific credibility, policy/social relevance and practical monitoring and data requirements.

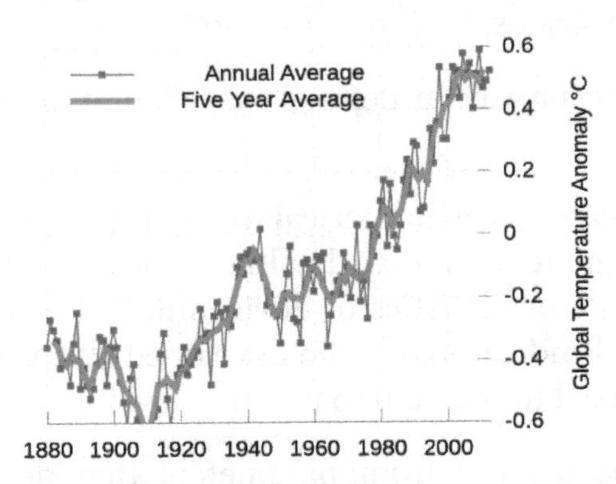

An example of an environmental indicator: Trend in global temperature anomalies of the last 150 years as an indicator of climate change

Environmental indicators are used by governments, non-government organizations, community groups and research institutions to see if environmental objectives are being met, to communicate the state of the environment to the general public and decision makers and as a diagnostic tool through detecting trends in the environment.

Environmental indicators can be measured and reported at different scales. For example, a town may track air quality along with water quality and count the number of rare species of birds to estimate the health of the environment in their area. Indicators are developed for specific ecosystems, such as the Great-Lakes in North America. National governments use environmental indicators to show status and trends with respect to environmental issues of importance to their citizens.

Use in Assessment

Some have attempted to monitor and assess the state of the planet using indicators.

Lester Brown of the Earth Policy Institute, has said: —

> Environmentally, the world is in an overshoot mode. If we use environmental indicators to evaluate our situation, then the global decline of the economy's natural support systems - the environmental decline that will lead to economic decline and social collapse- is well underway.

Environmental indicators are also used by companies in the framework of an Environmental management system. The EU Eco-Management and Audit Scheme provides core indicators or Performance Indicator (KPIs) with which registered organizations can measure their environmental performance and monitor their continual environmental improvement against set targets.

Audience

The types of indicators selected or developed should be partially based on who will be using the information from the indicators. There are generally three possible audiences to consider, each with different information needs. These audiences are: 1) technical experts and science advisors, 2) policy-makers, decision makers and resource managers, and 3) general public and media.

The technical experts and scientists will be interested in detailed and complex indicators. These indicators should have scientific validity, sensitivity, responsiveness and have data available on past conditions. The audience that includes policy-makers and resource managers will be concerned with using indicators that are directly related to evaluating policies and objectives. They require their indicators to be sensitive, responsive and have historical data available like the technical audience, but they are also looking for indicators that are cost-effective and have meaning for public awareness. Finally, the general public responds to indicators that have clear and simple messages and are meaningful to them, such as the UV index and the air quality index.

Indicator Systems and Communicating Them

Individual indicators are designed to translate complex information in a concise and easily understood manner in order to represent a particular phenomenon (e.g. ambient air quality). In contrast, indicator systems (or collections of indicators), when seen as a whole are meant to provide an assessment of the full environment domain or a major subset of it (e.g. forests).

Some indicator systems have evolved to include many indicators and require a certain level of knowledge and expertise in various disciplines to fully grasp. A number of methods have been devised in the recent past to boil down this information and allow for rapid consumption by those who do not have the time or the expertise to analyse the full set of indicators. In general these methods can be categorized as *numerical aggregation* (e.g. indices), *short selections of indicators* (e.g. core set or headline indicators), *short visual assessments* (e.g. arrows, traffic signals), and *compelling presentations* (e.g. maps or the dashboard of sustainability). Many prominent environmental indicator systems have adjusted their indicator systems to include or report solely on a limited "indicator set" (e.g. the OECD's "Key Environmental Indicators" and the "Canadian Environmental Sustainability Indicators")

Biosignature

A biosignature is any substance – such as an element, isotope, molecule, or phenomenon – that provides scientific evidence of past or present life. Measurable attributes of life include its complex physical and chemical structures and also its utilization of free energy and the production of biomass and wastes. Due to its unique characteristics, a biosignature can be interpreted as having been produced by living organisms; however, it is important that they not be considered definitive because there is no way of knowing in advance which ones are universal to life and which ones are unique to the peculiar circumstances of life on Earth. Nonetheless, life forms are known to shed unique chemicals, including DNA, into the environment as evidence of their presence in a particular location.

In Geomicrobiology

The ancient record on Earth provides an opportunity to see what geochemical signatures are produced by microbial life and how these signatures are preserved over geologic time. Some related disciplines such as geochemistry, geobiology, and geomicrobiology often use biosignatures to determine if living organisms are or were present in a sample. These possible biosignatures include: (a) microfossils and stromatolites; (b) molecular structures (biomarkers) and isotopic compositions of carbon, nitrogen and hydrogen in organic matter; (c) multiple sulfur and oxygen isotope ratios of minerals; and (d) abundance relationships and isotopic compositions of redox sensitive metals (e.g., Fe, Mo, Cr, and rare earth elements).

Electron micrograph of microfossils from a sediment core obtained by the Deep Sea Drilling Program

For example, the particular fatty acids measured in a sample can indicate which types of bacteria and archaea live in that environment. Another example are the long-chain fatty alcohols with more than 23 atoms that are produced by planktonic bacteria. When used in this sense, geochemists often prefer the term biomarker. Another example is the presence of straight-chain lipids in the form of alkanes, alcohols an fatty acids with 20-36 carbon atoms in soils or sediments. Peat deposits are an indication of originating from the epicuticular wax of higher plants.

Life processes may produce a range of biosignatures such as nucleic acids, lipids, proteins, amino acids, kerogen-like material and various morphological features that are detectable in rocks and sediments. Microbes often interact with geochemical processes, leaving features in the rock record indicative of biosignatures. For example, bacterial micrometer-sized pores in carbonate rocks resemble inclusions under transmitted light, but have distinct size, shapes and patterns (swirling or dendritic) and are distributed differently from common fluid inclusions. A potential biosignature is a phenomenon that *may* have been produced by life, but for which alternate abiotic origins may also be possible.

In Astrobiology

Astrobiological exploration is founded upon the premise that biosignatures encountered in space will be recognizable as extraterrestrial life. The usefulness of a biosignature is determined, not only by the probability of life creating it, but also by the improbability of nonbiological (abiotic) processes producing it. An example of such a biosignature might be complex organic molecules and/or structures whose formation is virtually unachievable in the absence of life. For example, some categories of biosignatures can include the following: cellular and extracellular morphologies, biomolecules in rocks, bio-organic molecular structures, chirality, biogenic minerals, biogenic stable isotope patterns in minerals and organic compounds, atmospheric gases, and remotely detectable features on planetary surfaces, such as photosynthetic pigments, etc.

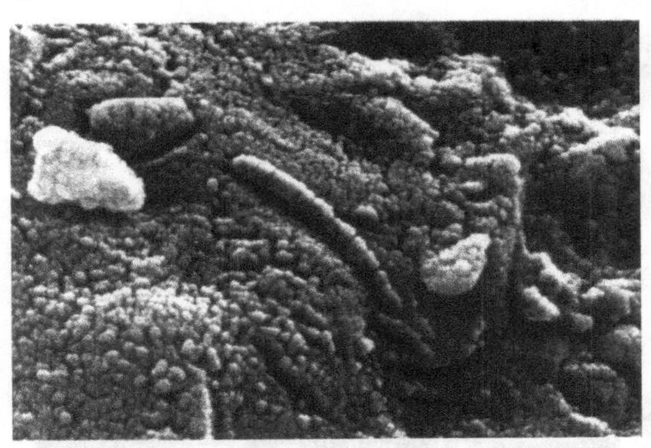

Some researchers suggested that these microscopic structures on the Martian ALH84001 meteorite could be fossilized bacteria.

Biosignatures need not be chemical, however, and can also be suggested by a distinctive magnetic biosignature. Another possible biosignature might be morphology since the shape and size of certain objects may potentially indicate the presence of past or present life. For example, microscopic magnetite crystals in the Martian meteorite ALH84001 were the longest-debated of several potential biosignatures in that specimen because it was believed until recently that only bacteria could create crystals of their specific shape. However, anomalous features discovered that are "possible biosignatures" for life forms would be investigated as well. Such features constitute a working hypothesis, not a confirmation of detection of life. Concluding that evidence of an extraterrestrial life form (past or present) has been discovered, requires proving that a possible biosignature was produced by the activities or remains of life. For example, the possible biomineral studied in the Martian ALH84001 meteorite includes putative microbial fossils, tiny rock-like structures whose shape was a potential biosignature because it resembled known bacteria. Most scientists ultimately concluded that these were far too small to be fossilized cells. A consensus that has emerged from these discussions, and is now seen as a critical requirement, is the demand for further lines of evidence in addition to any morphological data that supports such extraordinary claims.

Scientific observations include the possible identification of biosignatures through indirect observation. For example, electromagnetic information through infrared radiation telescopes, radio-telescopes, space telescopes, etc. From this discipline, the hypothetical electromagnetic radio signatures that SETI scans for would be a biosignature, since a message from intelligent aliens would certainly demonstrate the existence of extraterrestrial life.

On Mars, surface oxidants and UV radiation will have altered or destroyed organic molecules at or near the surface. One issue that may add ambiguity in such a search is the fact that, throughout Martian history, abiogenic organic-rich chondritic meteorites have undoubtedly rained upon the Martian surface. At the same time, strong oxidants in Martian soil along with exposure to ionizing radiation might alter or destroy molecular signatures from meteorites or organisms. An alternative approach would be to seek concentrations of buried crystalline minerals, such as clays and evaporites, which may protect organic matter from the destructive effects of ionizing radiation and strong oxidants. The search for Martian biosignatures has become more promising due to the discovery that surface and near-surface aqueous environments existed on Mars at the same time when biological organic matter was being preserved in ancient aqueous sediments on Earth.

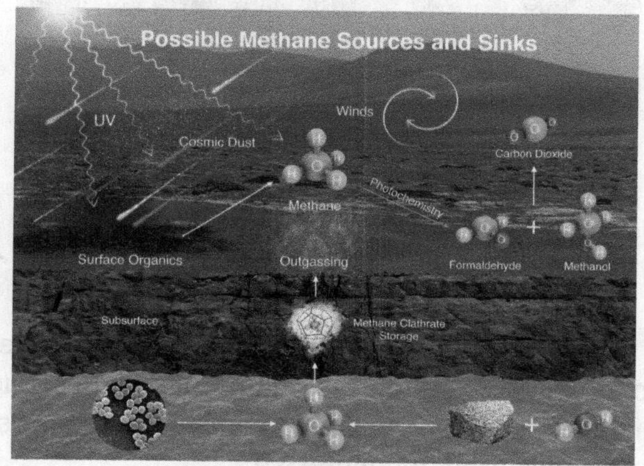

Methane (CH_4) on Mars - potential sources and sinks.

Atmosphere

Over billions of years, the processes of life on a planet would result in a mixture of chemicals unlike anything that could form in an ordinary chemical equilibrium. For example, large amounts of oxygen and small amounts of methane are generated by life on Earth.

Also, an exoplanet's color —or reflectance spectrum— might give away the presence of vast colonies of life forms at its surface.

The presence of methane in the atmosphere of Mars indicates that there must be an active source on the planet, as it is an unstable gas. Furthermore, current photochemical models cannot explain the presence of methane in the atmosphere of Mars and its reported rapid variations in space and time. Neither its fast appearance nor disappearance can be explained yet. To rule out a biogenic origin for the methane, a future probe or lander hosting a mass spectrometer will be needed, as the isotopic proportions of carbon-12 to carbon-14 in methane could distinguish between a biogenic and non-biogenic origin. In June, 2012, scientists reported that measuring the ratio of hydrogen and methane levels on Mars may help determine the likelihood of life on Mars. According to the scientists,

"...low H_2/CH_4 ratios (less than approximately 40) indicate that life is likely present and active." Other scientists have recently reported methods of detecting hydrogen and methane in extraterrestrial atmospheres. The planned ExoMars Trace Gas Orbiter, launched in March 2016 to Mars, will study atmospheric trace gases and will attempt to characterize potential biochemical and geochemical processes at work.

The Viking Missions to Mars

Carl Sagan with a model of the Viking lander

The Viking missions to Mars in the 1970s conducted the first experiments which were explicitly designed to look for biosignatures on another planet. Each of the two Viking landers carried three life-detection experiments which looked for signs of metabolism; however, the results were declared 'inconclusive'.

Mars Science Laboratory

The *Curiosity* rover from the Mars Science Laboratory mission, is currently assessing the potential past and present habitability of the Martian environment and is attempting to detect biosignatures on the surface of Mars. Considering the MSL instrument payload package, the following classes of biosignatures are within the MSL detection window: organism morphologies (cells, body fossils, casts), biofabrics (including microbial mats), diagnostic organic molecules, isotopic signatures, evidence of biomineralization and bioalteration, spatial patterns in chemistry, and biogenic gases. Of these, biogenic organic molecules and biogenic atmospheric gases are considered the most definitive and most readily detectable by MSL. The *Curiosity* rover targets outcrops to maximize the probability of detecting 'fossilized' organic matter preserved in sedimentary deposits.

On January 24, 2014, NASA reported that current studies by the *Curiosity* and *Opportunity* rovers on the planet Mars will now be searching for evidence of ancient life, including a biosphere based on autotrophic, chemotrophic and/or chemolithoautotrophic microorganisms, as well as ancient water, including fluvio-lacustrine environments (plains related to ancient rivers or lakes) that may have been habitable. The search for evidence of habitability, taphonomy (related to fossils), and organic carbon on the planet Mars is now a primary NASA objective.

ExoMars

The 2016 Trace Gas Orbiter (TGO) will be a Mars telecommunications orbiter and atmospheric gas analyzer mission. It will deliver the ExoMars EDM lander and then proceed to map the sources of methane on Mars and other gases, and in doing so, help select the landing site for the ExoMars rover to be launched on 2018. The primary objective of the 2018 ExoMars rover mission is the search for biosignatures on the surface and subsurface by using a drill able to collect samples down to a depth of 2 metres (6.6 ft).

Copepod

Copepods are a group of small crustaceans found in the sea and nearly every freshwater habitat. Some species are planktonic (drifting in sea waters), some are benthic (living on the ocean floor), and some continental species may live in limnoterrestrial habitats and other wet terrestrial places, such as swamps, under leaf fall in wet forests, bogs, springs, ephemeral ponds, and puddles, damp moss, or water-filled recesses (phytotelmata) of plants such as bromeliads and pitcher plants. Many live underground in marine and freshwater caves, sinkholes, or stream beds. Copepods are sometimes used as biodiversity indicators.

As with other crustaceans, copepods have a larval form. For copepods, the eggs hatches into a naplius form, with a head and a tail but no true thorax or abdomen. The larva molts several times until it resembles the adult and then, after more molts, achieves adult development. The naplius form is so different from the adult form that it was once thought to be a separate species.

Classification and Diversity

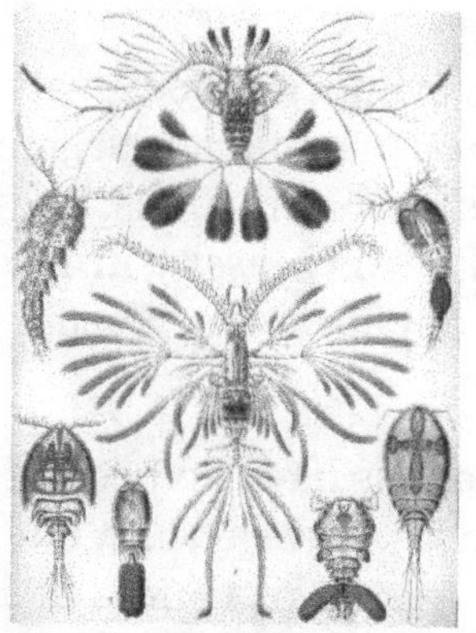

Copepods from Ernst Haeckel's *Kunstformen der Natur*

Copepods form a subclass belonging to the subphylum Crustacea (crustaceans); they are divided into ten orders. Some 13,000 species of copepods are known, and 2,800 of them live in fresh water.

Characteristics

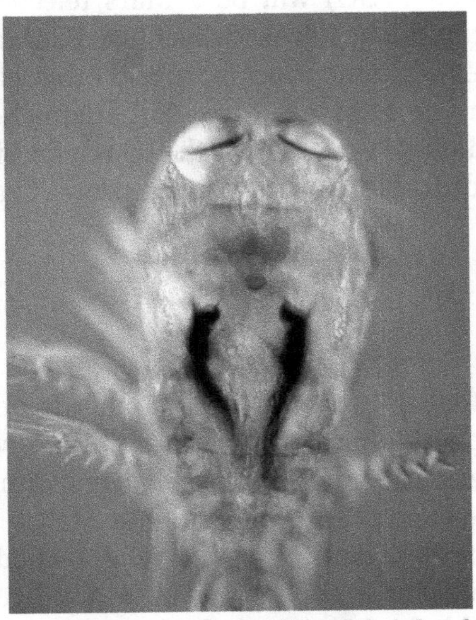

Most copepods have a single naupliar eye in the middle of their head, but copepods of the genus *Corycaeus* possess two large cuticular lenses paired to form a telescope.

Copepods vary considerably, but can typically be 1 to 2 mm (0.04 to 0.08 in) long, with a teardrop-shaped body and large antennae. Although like other crustaceans, they have an armoured exoskeleton, they are so small that in most species, this thin armour, and the entire body, is almost totally transparent. Some polar copepods reach 1 cm (0.39 in). Most copepods have a single median compound eye, usually bright red and in the centre of the transparent head; subterranean species may be eyeless. Like other crustaceans, copepods possess two pairs of antennae; the first pair is often long and conspicuous.

Copepods typically have a short, cylindrical body, with a rounded or beaked head. The head is fused with the first one or two thoracic segments, while the remainder of the thorax has three to five segments, each with limbs. The first pair of thoracic appendages is modified to form maxillipeds, which assist in feeding. The abdomen is typically narrower than the thorax, and contains five segments without any appendages, except for some tail-like "rami" at the tip.

Because of their small size, copepods have no need of any heart or circulatory system (the members of the order Calanoida have a heart, but no blood vessels), and most also lack gills. Instead, they absorb oxygen directly into their bodies. Their excretory system consists of maxillary glands.

Behavior

The second pair of cephalic appendages in free-living copepods is usually the main time-averaged source of propulsion, beating like oars to pull the animal through the water. However, different groups have different modes of feeding and locomotion, ranging from almost immotile for several minutes (e.g. some harpacticoid copepods) to intermittent motion (e.g., some cyclopoid copepods) and continuous displacements with some escape reactions (e.g. most calanoid copepods.)

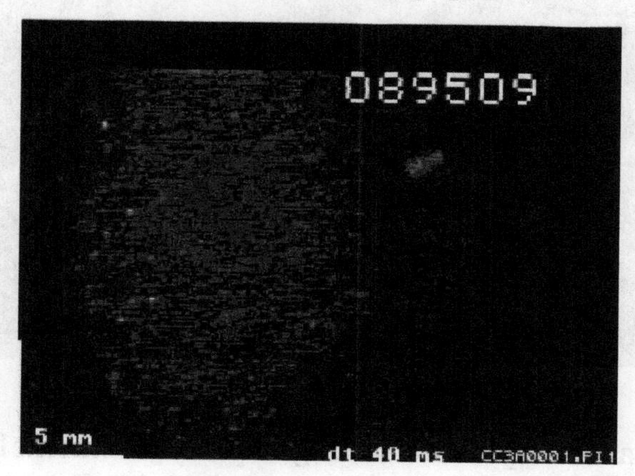

Slow-motion macrophotography video (50%), taken using ecoSCOPE, of juvenile Atlantic herring (38 mm) feeding on copepods – the fish approach from below and catch each copepod individually. In the middle of the image, a copepod escapes successfully to the left.

Some copepods have extremely fast escape responses when a predator is sensed and can jump with high speed over a few millimetres. Many species have neurons surrounded by myelin (for increased conduction speed), which is very rare among invertebrates (other examples are some annelids and malacostracan crustaceans like palaemonid shrimp and penaeids). Even rarer, the myelin is highly organized, resembling the well-organized wrapping found in vertebrates (Gnathostomata). Despite their fast escape response, copepods are successfully hunted by slow-swimming seahorses, which approach their prey so gradually, it senses no turbulence, then suck the copepod into their snout too suddenly for the copepod to escape.

Finding a mate in the three-dimensional space of open water is challenging. Some copepod females solve the problem by emitting pheromones, which leave a trail in the water that the male can follow.

Diet

Copepods feed directly on phytoplankton, catching cells singly. Some of the larger species are predators of their smaller relatives. Many benthic copepods eat organic detritus or the bacteria that grow in it, and their mouth parts are adapted for scraping and biting. Herbivorous copepods, particularly those in rich, cold seas, store up energy from their food as oil droplets while they feed in the spring and summer on plankton blooms. These droplets may take up over half of the volume of their bodies in polar species. Many copepods (e.g., fish lice like the Siphonostomatoida) are parasites, and feed on their host organisms.

Life Cycle

Egg sac of a copepod

Copepods are a holoplankton species, meaning they stay planktonic for all of their lifecycle. During mating, the male copepod grips the female with his first pair of antennae, which is sometimes modified for this purpose. The male then produces an adhesive package of sperm and transfers it to the female's genital opening with his thoracic limbs. Eggs are sometimes laid directly into the water, but many species enclose them within a sac attached to the female's body until they hatch. In some pond-dwelling species, the eggs have a tough shell and can lie dormant for extended periods if the pond dries up.

Eggs hatch into nauplius larvae, which consist of a head with a small tail, but no thorax or true abdomen. The nauplius moults five or six times, before emerging as a "copepodid larva". This stage resembles the adult, but has a simple, unsegmented abdomen and only three pairs of thoracic limbs. After a further five moults, the copepod takes on the adult form. The entire process from hatching to adulthood can take a week to a year, depending on the species.

Ecology

Planktonic copepods are important to global ecology and the carbon cycle. They are usually the dominant members of the zooplankton, and are major food organisms for small fish such as the dragonet, banded killifish, whales, seabirds, alaska pollock and other crustaceans such as krill in the ocean and in fresh water. Some scientists say they form the largest animal biomass on earth. Copepods compete for this title with Antarctic krill (*Euphausia superba*). *C. glacialis* inhabits the edge of the Arctic icepack, where they alone comprise up to 80% of zooplankton biomass. They bloom as the ice recedes each spring. The ongoing large reductions in the annual minimum of recent years may force them to compete in the open ocean with the much-less nourishing *C. finmarchicus*, which is spreading from the North Sea and the Norwegian Sea into the Barents Sea.

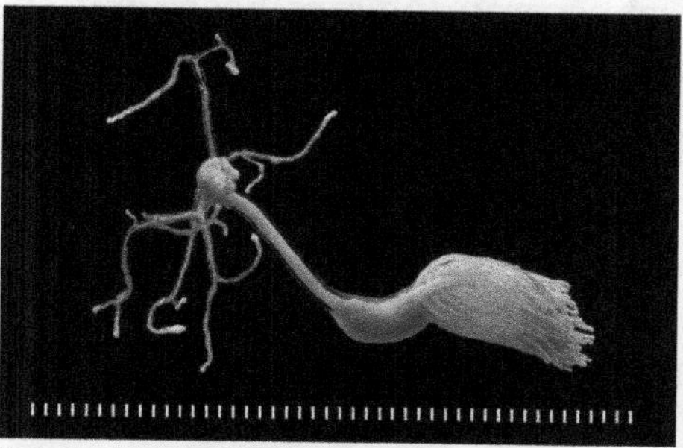

Lernaeolophus sultanus (Pennellidae), parasite of the fish *Pristipomoides filamentosus*, scale: each division = 1 mm

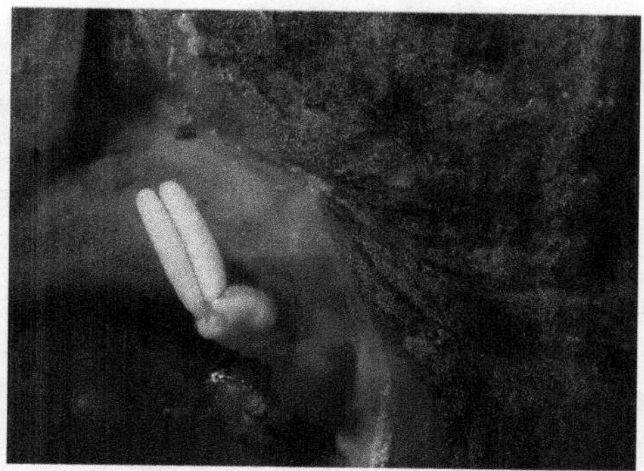

Acanthochondria cornuta, an ectoparasite on flounder in the North Sea

Because of their smaller size and relatively faster growth rates, and because they are more evenly distributed throughout more of the world's oceans, copepods almost certainly contribute far more to the secondary productivity of the world's oceans, and to the global ocean carbon sink than krill, and perhaps more than all other groups of organisms together. The surface layers of the oceans are currently believed to be the world's largest carbon sink, absorbing about 2 billion tons of carbon a year, the equivalent to perhaps a third of human carbon emissions, thus reducing their impact. Many planktonic copepods feed near the surface at night, then sink (by changing oils into more dense fats) into deeper water during the day to avoid visual predators. Their moulted exoskeletons, faecal pellets, and respiration at depth all bring carbon to the deep sea.

About half of the estimated 13,000 described species of copepods are parasitic and have strongly modified bodies. They attach themselves to bony fish, sharks, marine mammals, and many kinds of invertebrates such as molluscs, tunicates, or corals. They live as endo- or ectoparasites on fish or invertebrates in fresh water as well as in marine environments.

Copepods as Parasitic Hosts

In addition to being parasites themselves, copepods are subject to parasitic infection. The most common parasite is the marine dinoflagellate, *Blastodinium* spp., which are gut parasites of many copepod species. Currently 12 species of *Blastodinium* are described, the majority of which were discovered in the Mediterranean Sea. Most *Blastodinium* species infect several different hosts, but species-specific infection of copepods does occur. Generally, adult copepod females and juveniles are infected.

During the naupliar stage, the copepod host ingests the unicellular dinospore of the parasite. The dinospore is not digested and continues to grow inside the intestinal lumen of the copepod. Eventually, the parasite divides into a multicellular arrangement called a trophont. This trophont is considered parasitic, contains thousands of cells, and can be several hundred micrometers in length. The trophont is greenish to brownish in color as a result of well-defined chloroplasts. At maturity, the trophont ruptures and *Blastodinium* spp. are released from the copepod anus as free dinospore cells. Not much is known about the dinospore stage of *Blastodinium* and its ability to persist outside of the copepod host in relatively high abundances.

The copepod, *Calanus finmarchicus*, which dominates the northeastern Atlantic coast, has been shown to be greatly infected by this parasite. A 2014 study in this region found up to 58% of collected *C. finmarchicus* females to be infected. In this study, *Blastodinium*-infected females had no measurable feeding rate over a 24-hour period. This is compared to uninfected females which, on average, ate 2.93×10^4 cells $copepod^{-1}$ d^{-1}. *Blastodinium*-infected females of *C. finmarchicus* exhibited characteristic signs of starvation including decreased respiration, fecundity, and fecal pellet production. Though photosynthetic, *Blastodinium* spp. procure most of their energy from organic material in the copepod gut, thus attributing to host starvation. Underdeveloped or disintegrated ovaries, as well as decreased fecal pellet size are a direct result of starvation in female copepods. Infection from *Blastodinium* spp. could have serious ramifications on the success of copepod species and the function of entire marine ecosystems. Parasitism via *Blastodinium* spp.' is not lethal, but has negative impacts on copepod physiology, which in turn may alter marine biogeochemical cycles.

Copepods also host *Dracunculus medinensis*, the Guinea worm nematode that causes dracunculiasis disease in humans. This disease may be close to being eradicated through efforts at the U.S. Centers for Disease Control and Prevention and the World Health Organization.

Practical Aspects

Copepods in Marine Aquaria

Live copepods are used in the saltwater aquarium hobby as a food source and are generally considered beneficial in most reef tanks. They are scavengers and also may feed on algae, including coralline algae. Live copepods are popular among hobbyists who are attempting to keep particularly difficult species such as the mandarin dragonet or scooter blenny. They are also popular to hobbyists who want to breed marine species in captivity. In a saltwater aquarium, copepods are typically stocked in the refugium.

Water Supplies

Copepods are sometimes found in public main water supplies, especially systems where the water is not filtered, such as New York City, Boston, and San Francisco. This is not usually a problem in treated water supplies. In some tropical countries, such as Peru and Bangladesh, a correlation has been found between copepods presence and cholera in untreated water, because the cholera bacteria attach to the surfaces of planktonic animals. The larvae of the guinea worm must develop within a copepod's digestive tract before being transmitted to humans. The risk of infection with these diseases can be reduced by filtering out the copepods (and other matter), for example with a cloth filter.

Copepods have been used successfully in Vietnam to control disease-bearing mosquitoes such as *Aedes aegypti* that transmit dengue fever and other human parasitic diseases.

The copepods can be added to water-storage containers where the mosquitoes breed. Copepods, primarily of the genera *Mesocyclops* and *Macrocyclops* (such as *Macrocyclops albidus*), can survive for periods of months in the containers, if the containers are not completely drained by their users. They attack, kill, and eat the younger first- and second-instar larvae of the mosquitoes. This biological control method is complemented by community trash removal and recycling to eliminate other possible mosquito-breeding sites. Because the water in these containers is drawn from uncontaminated sources such as rainfall, the risk of contamination by cholera bacteria is small, and in fact no cases of cholera have been linked to copepods introduced into water-storage containers. Trials using copepods to control container-breeding mosquitoes are underway in several other countries, including Thailand and the southern United States. The method, though, would be very ill-advised in areas where the guinea worm is endemic.

The presence of copepods in the New York City water supply system has caused problems for some Jewish people who observe Kashrut. Copepods, being crustaceans, are not kosher, nor are they small enough to be ignored as nonfood microscopic organisms, since some specimens can be seen with the naked eye. When a group of rabbis in Brooklyn, New York discovered the copepods in the summer of 2004, they triggered such enormous debate in rabbinic circles that some observant Jews felt compelled to buy and install filters for their water. The water was ultimately ruled kosher by posek Yisrael Belsky.

Aquatic Biomonitoring

Aquatic biomonitoring is the science of inferring the ecological condition of rivers, lakes, streams, and wetlands by examining the organisms that live there. While aquatic biomonitoring is the most common form of such biomonitoring, any ecosystem can be studied in this manner.

A biosurvey on the North Toe River. North Carolina

Biomonitoring typically takes different approaches:

- *Bioassays*, where test organisms are exposed to an environment to see if mutations or deaths occur. Typical organisms used in bioassays are fish, water fleas (Daphnia), and frogs.
- *Community assessments*, also called *biosurveys,* where an entire community of organisms is sampled, to see what types of taxa remain. In aquatic ecosystems, these assessments often focus on invertebrates, algae, macrophytes (aquatic plants), fish, or amphibians. Rarely, other large vertebrates (reptiles, birds, and mammals) are considered as well.
- *Online biomonitoring devices*, using the ability of animals to permanently taste their environment. Different types of animals are used for that purpose either under lab or field conditions. The use of valve opening/closing activity of clams is one of the possible ways to monitor *in-situ* the quality of freshwater and coastal waters.

Aquatic invertebrates have the longest history of use in biomonitoring programs. In typical unpolluted temperate streams of Europe and North America, certain insect taxa predominate. Mayflies (Ephemeroptera), caddisflies (Trichoptera), and stoneflies (Plecoptera) are the most common insects in these undisturbed streams. In rivers disturbed by urbanization, agriculture, forestry, and other perturbations, flies (Diptera), and especially midges (family Chironomidae) predominate. Aquatic invertebrates are responsive to climate change.

Bioassay

Bioassay (commonly used shorthand for biological assay or assessment), or biological standardization is a type of scientific experiment. A bioassay involves the use of live animal or plant (*in vivo*) or tissue or cell (*in vitro*) to determine the biological activity of a substance, such as a hormone or drug. Bioassays are typically conducted to measure the effects of a substance on a living organism and are essential in the development of new drugs and in monitoring environmental pollutants. Both are procedures by which the potency or the nature of a substance is estimated by studying its effects on living matter. A bioassay can also be used to determine the concentration of a particular constitution of a mixture that may cause harmful effects on organisms or the environment.

Use

Bioassays are procedures that can determine the concentration or purity or biological activity of a substance such as vitamin, hormone or plant growth factor by measuring the effect on an organism, tissue, cells, enzyme or receptor. Bioassays may be qualitative or quantitative. Qualitative bioassays are used for assessing the physical effects of a substance that may not be quantified, such as seeds fail to germinate or develop abnormally deformity. An example of a qualitative bioassay includes Arnold Adolph Berthold's famous experiment on castrated chickens. This analysis found that by removing the testicles of a chicken, it would not develop into a rooster because the endocrine signals necessary for this process were not available. Quantitative bioassays involve estimation of the dose-response curve, how the response changes with increasing dose. That dose-response relation allows estimation of the dose or concentration of a substance associated with a specific biological response, such as the LC50 (concentration killing 50% of the exposed organisms). Quantitative bioassays are typically analyzed using the methods of biostatistics. For more information Look up Basic and Clinical Pharmacology by Bertram G. Katzung.

Definition

"The determination of the relative strength of a substance (e.g., a drug or hormone or toxicant) by comparing its effect on a test organism with that of a standard preparation" is called bioassay.

Similarly, Bioassay is a method of developing toxicological information on organisms whose physiology is considered similar to the organisms of direct concern to a known level of toxic chemical compound in an environmentally concerned chamber.

Purpose

1. Measurement of the pharmacological activity of new or chemically undefined substances
2. Investigation of the function of endogenous mediators
3. Determination of the side-effect profile, including the degree of drug toxicity
4. Measurement of the concentration of known substances (alternatives to the use of whole animals have made this use obsolete)
5. Assessing the amount of pollutants being released by a particular source, such as wastewater or urban runoff.
6. Determining the specificity of certain enzymes to certain substrates.

Types

Bioassays are of two types:

Quantal

A quantal assay involves an "all or none response".

Graded

Graded assays are based on the observation that there is a proportionate increase in the observed response following an increase in the concentration or dose. The parameters employed in such bioassays are based on the nature of the effect the substance is expected to produce. For example: contraction of smooth muscle preparation for assaying histamine or the study of blood pressure response in case of adrenaline.

A graded bioassay can be performed by employing any of the below-mentioned techniques. The choice of procedure depends on:

1. the precision of the assay required
2. the quantity of the sample substance available
3. the availability of the experimental animals.

Techniques

1. Matching Bioassay
2. Interpolation Method
3. Bracketing Method

4. Multiple Point Bioassay (i.e.-Three-point, Four-point and Six Point Bioassay)
5. divided bioassy

Matching Bioassay: It is the simplest type of the bioassay. In this type of bioassay, response of the test substance taken first and the observed response is tried to match with the standard response. Several responses of the standard drug are recorded till a close matching point to that of the test substance is observed. A corresponding concentration is thus calculated. This assay is applied when the sample size is too small. Since the assay does not involve the recording of concentration response curve, the sensitivity of the preparation is not taken into consideration. Therefore, precision and reliability is not very good.

Interpolation bioassay: Bioassays are conducted by determining the amount of preparation of unknown potency required to produce a definite effect on suitable test animals or organs or tissue under standard conditions. This effect is compared with that of a standard. Thus the amount of the test substance required to produce the same biological effect as a given quantity the unit of a standard preparation is compared and the potency of the unknown is expressed as a % of that of the standard by employing a simple formula.

Many times, a reliable result cannot be obtained using this calculation. Therefore it may be necessary to adopt more precise methods of calculating potency based upon observations of relative, but not necessarily equal effects, likewise, statistical methods may also be employed. The data (obtained from either of assay techniques used) on which bioassay are based may be classified as quantal or graded response. Both these depend ultimately on plotting or making assumption concerning the form of DRC.

Environmental Bioassays

Environmental bioassays are generally a broad-range survey of toxicity. A toxicity identification evaluation is conducted to determine what the relevant toxicants are. Although bioassays are beneficial in determining the biological activity within an organism, they can often be time-consuming and laborious. Organism-specific factors may result in data that are not applicable to others in that species. For these reasons, other biological techniques are often employed, including radioimmunoassays.

Water pollution control requirements in the United States require some industrial dischargers and municipal sewage treatment plants to conduct bioassays. These procedures, called whole effluent toxicity tests, include acute toxicity tests as well as chronic test methods. The methods involve exposing living aquatic organisms to samples of wastewater for a specific length of time. Another example is the bioassay ECOTOX, which uses the microalgae *Euglena gracilis* to test the toxicity of water samples.

Photosynthesis

Photosynthesis is a process used by plants and other organisms to convert light energy into chemical energy that can later be released to fuel the organisms' activities (energy transformation). This chemical energy is stored in carbohydrate molecules, such as sugars, which are synthesized from carbon dioxide and water – hence the name *photosynthesis*. In most cases, oxygen is also released as a waste product. Most plants, most algae, and cyanobacteria perform photosynthesis; such organisms are called photoautotrophs. Photosynthesis is largely responsible for producing and maintaining the oxygen content of the Earth's atmosphere, and supplies all of the organic compounds and most of the energy necessary for life on Earth.

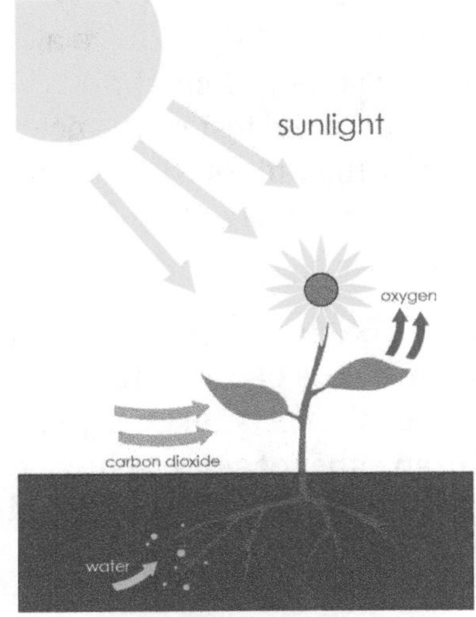

Schematic of photosynthesis in plants. The carbohydrates produced are stored in or used by the plant.

$$\underset{\text{Carbon dioxide}}{6CO_2} + \underset{\text{Water}}{6H_2O} \xrightarrow{\text{Light}} \underset{\text{Sugar}}{C_6H_{12}O_6} + \underset{\text{Oxygen}}{6O_2}$$

Overall equation for the type of photosynthesis that occurs in plants

Composite image showing the global distribution of photosynthesis, including both oceanic phytoplankton and terrestrial vegetation. Dark red and blue-green indicate regions of high photosynthetic activity in the ocean and on land, respectively.

Although photosynthesis is performed differently by different species, the process always begins when energy from light is absorbed by proteins called reaction centres that contain green chlorophyll pigments. In plants, these proteins are held inside organelles called chloroplasts, which are most abundant in leaf cells, while in bacteria they are embedded in the plasma membrane. In these light-dependent reactions, some energy is used to strip electrons from suitable substances, such as water, producing oxygen gas. The hydrogen freed by the splitting of water is used in the creation of two further compounds that act as an immediate energy storage means: reduced nicotinamide adenine dinucleotide phosphate (NADPH) and adenosine triphosphate (ATP), the "energy currency" of cells.

In plants, algae and cyanobacteria, long-term energy storage in the form of sugars is produced by a subsequent sequence of light-independent reactions called the Calvin cycle; some bacteria use different mechanisms, such as the reverse Krebs cycle, to achieve the same end. In the Calvin cycle, atmospheric carbon dioxide is incorporated into already existing organic carbon compounds, such as ribulose bisphosphate (RuBP). Using the ATP and NADPH produced by the light-dependent reactions, the resulting compounds are then reduced and removed to form further carbohydrates, such as glucose.

The first photosynthetic organisms probably evolved early in the evolutionary history of life and most likely used reducing agents such as hydrogen or hydrogen sulfide, rather than water, as sources of electrons. Cyanobacteria appeared later; the excess oxygen they produced contributed directly to the oxygenation of the Earth, which rendered the evolution of complex life possible. Today, the average rate of energy capture by photosynthesis globally is approximately 130 terawatts, which is about three times the current power consumption of human civilization. Photosynthetic organisms also convert around 100–115 thousand million metric tonnes of carbon into biomass per year.

Overview

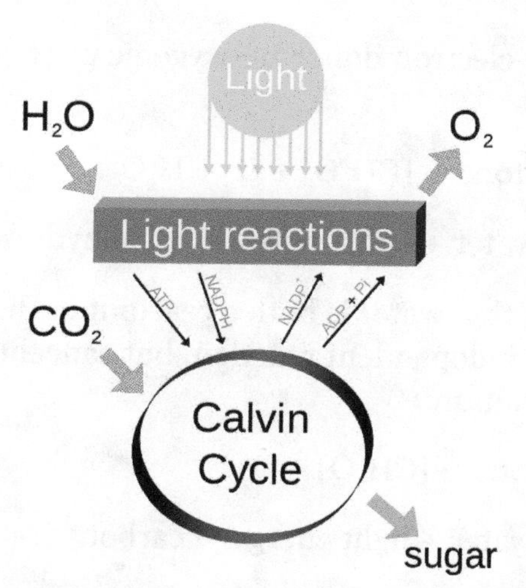

Photosynthesis changes sunlight into chemical energy, splits water to liberate O_2, and fixes CO_2 into sugar.

Photosynthetic organisms are photoautotrophs, which means that they are able to synthesize food directly from carbon dioxide and water using energy from light. However, not all organisms that use light as a source of energy carry out photosynthesis; photoheterotrophs use organic compounds, rather than carbon dioxide, as a source of carbon. In plants, algae, and cyanobacteria, photosynthesis releases oxygen. This is called *oxygenic photosynthesis* and is by far the most common type of photosynthesis used by living organisms. Although there are some differences between oxygenic photosynthesis in plants, algae, and cyanobacteria, the overall process is quite similar in these organisms. There are also many varieties of anoxygenic photosynthesis, used mostly by certain types of bacteria, which consume carbon dioxide but do not release oxygen.

Carbon dioxide is converted into sugars in a process called carbon fixation. Carbon fixation is an endothermic redox reaction, so photosynthesis needs to supply both a source of energy to drive this process, and the electrons needed to convert carbon dioxide into a carbohydrate via a reduction reaction. The addition of electrons to a chemical species is called reduction. In general outline and in effect, photosynthesis is the opposite of cellular respiration, in which glucose and other compounds are oxidized to produce carbon dioxide and water, and to release chemical energy (an exothermic reaction) to drive the organism's metabolism. The two processes, reduction of carbon dioxide to carbohydrate and then later oxidation of the carbohydrate, are distinct: photosynthesis and cellular respiration take place through a different sequence of chemical reactions and in different cellular compartments.

The general equation for photosynthesis as first proposed by Cornelius van Niel is therefore:

$$CO_2 + 2H_2A + \text{photons} \rightarrow [CH_2O] + 2A + H_2O$$

carbon dioxide + electron donor + light energy → carbohydrate + oxidized electron donor + water

Since water is used as the electron donor in oxygenic photosynthesis, the equation for this process is:

$$CO_2 + 2H_2O + \text{photons} \rightarrow [CH_2O] + O_2 + H_2O$$

carbon dioxide + water + light energy → carbohydrate + oxygen + water

This equation emphasizes that water is both a reactant in the light-dependent reaction and a product of the light-independent reaction, but canceling *n* water molecules from each side gives the net equation:

$$CO_2 + H_2O + \text{photons} \rightarrow [CH_2O] + O_2$$

carbon dioxide + water + light energy → carbohydrate + oxygen

Other processes substitute other compounds (such as arsenite) for water in the electron-supply role; for example some microbes use sunlight to oxidize arsenite to arsenate: The equation for this reaction is:

CO_2 + (AsO3−3) + photons → (AsO3−4) + CO

carbon dioxide + arsenite + light energy → arsenate + carbon monoxide (used to build other compounds in subsequent reactions)

Photosynthesis occurs in two stages. In the first stage, *light-dependent reactions* or *light reactions* capture the energy of light and use it to make the energy-storage molecules ATP and NADPH. During the second stage, the *light-independent reactions* use these products to capture and reduce carbon dioxide.

Most organisms that utilize oxygenic photosynthesis use visible light for the light-dependent reactions, although at least three use shortwave infrared or, more specifically, far-red radiation.

Some organisms employ even more radical variants of photosynthesis. Some archea use a simpler method that employs a pigment similar to those used for vision in animals. The bacteriorhodopsin changes its configuration in response to sunlight, acting as a proton pump. This produces a proton gradient more directly, which is then converted to chemical energy. The process does not involve carbon dioxide fixation and does not release oxygen, and seems to have evolved separately from the more common types of photosynthesis.

Photosynthetic Membranes and Organelles

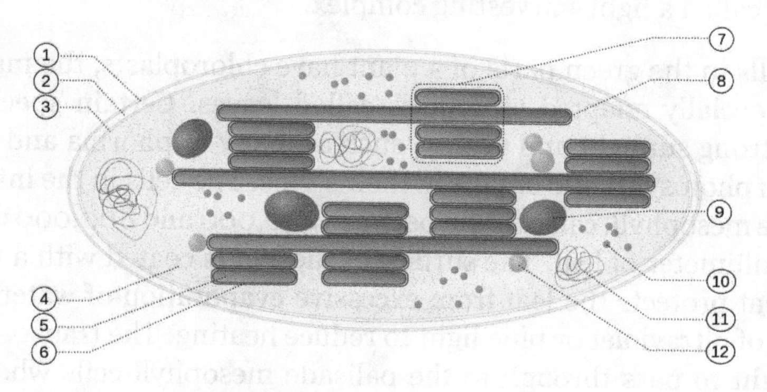

Chloroplast Ultrastructure:

1. outer membrane 2. intermembrane space 3. inner membrane (1+2+3: envelope) 4. stroma (aqueous fluid) 5. thylakoid lumen (inside of thylakoid) 6. thylakoid membrane 7. granum (stack of thylakoids)
8. thylakoid (lamella) 9. starch 10. ribosome 11. plastidial DNA 12. plastoglobule (drop of lipids)

In photosynthetic bacteria, the proteins that gather light for photosynthesis are embedded in cell membranes. In its simplest form, this involves the membrane surrounding the cell itself. However, the membrane may be tightly folded into cylindrical sheets called thylakoids, or bunched up into round vesicles called *intracytoplasmic membranes*. These structures can fill most of the interior of a cell, giving the membrane a very large surface area and therefore increasing the amount of light that the bacteria can absorb.

In plants and algae, photosynthesis takes place in organelles called chloroplasts. A typical plant cell contains about 10 to 100 chloroplasts. The chloroplast is enclosed by a membrane. This membrane is composed of a phospholipid inner membrane, a phospholipid outer membrane, and an intermembrane space. Enclosed by the membrane is an aqueous fluid called the stroma. Embedded within the stroma are stacks of thylakoids (grana), which are the site of photosynthesis. The thylakoids appear as flattened disks. The thylakoid itself is enclosed by the thylakoid membrane, and within the enclosed volume is a lumen or thylakoid space. Embedded in the thylakoid membrane are integral and peripheral membrane protein complexes of the photosynthetic system.

Plants absorb light primarily using the pigment chlorophyll. The green part of the light spectrum is not absorbed but is reflected which is the reason that most plants have a green color. Besides chlorophyll, plants also use pigments such as carotenes and xanthophylls. Algae also use chlorophyll, but various other pigments are present, such as phycocyanin, carotenes, and xanthophylls in green algae, phycoerythrin in red algae (rhodophytes) and fucoxanthin in brown algae and diatoms resulting in a wide variety of colors.

These pigments are embedded in plants and algae in complexes called antenna proteins. In such proteins, the pigments are arranged to work together. Such a combination of proteins is also called a light-harvesting complex.

Although all cells in the green parts of a plant have chloroplasts, the majority of those are found in specially adapted structures called leaves. Certain species adapted to conditions of strong sunlight and aridity, such as many Euphorbia and cactus species, have their main photosynthetic organs in their stems. The cells in the interior tissues of a leaf, called the mesophyll, can contain between 450,000 and 800,000 chloroplasts for every square millimeter of leaf. The surface of the leaf is coated with a water-resistant waxy cuticle that protects the leaf from excessive evaporation of water and decreases the absorption of ultraviolet or blue light to reduce heating. The transparent epidermis layer allows light to pass through to the palisade mesophyll cells where most of the photosynthesis takes place.

Light-dependent Reactions

In the light-dependent reactions, one molecule of the pigment chlorophyll absorbs one photon and loses one electron. This electron is passed to a modified form of chlorophyll called pheophytin, which passes the electron to a quinone molecule, starting the flow of electrons down an electron transport chain that leads to the ultimate reduction of NADP to NADPH. In addition, this creates a proton gradient (energy gradient) across the chloroplast membrane, which is used by ATP synthase in the synthesis of ATP. The chlorophyll molecule ultimately regains the electron it lost when a water molecule is split in a process called photolysis, which releases a dioxygen (O_2) molecule as a waste product.

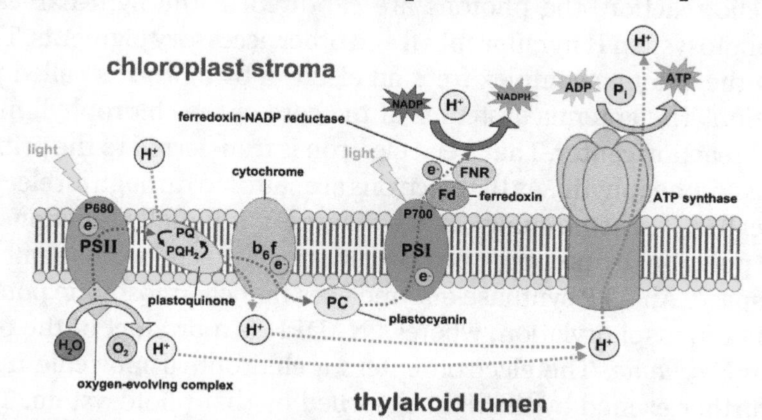

Light-dependent reactions of photosynthesis at the thylakoid membrane

The overall equation for the light-dependent reactions under the conditions of non-cyclic electron flow in green plants is:

$$2\ H_2O + 2\ NADP^+ + 3\ ADP + 3\ P_i + \text{light} \rightarrow 2\ NADPH + 2\ H^+ + 3\ ATP + O_2$$

Not all wavelengths of light can support photosynthesis. The photosynthetic action spectrum depends on the type of accessory pigments present. For example, in green plants, the action spectrum resembles the absorption spectrum for chlorophylls and carotenoids with peaks for violet-blue and red light. In red algae, the action spectrum is blue-green light, which allows these algae to use the blue end of the spectrum to grow in the deeper waters that filter out the longer wavelengths (red light) used by above ground green plants. The non-absorbed part of the light spectrum is what gives photosynthetic organisms their color (e.g., green plants, red algae, purple bacteria) and is the least effective for photosynthesis in the respective organisms.

Z Scheme

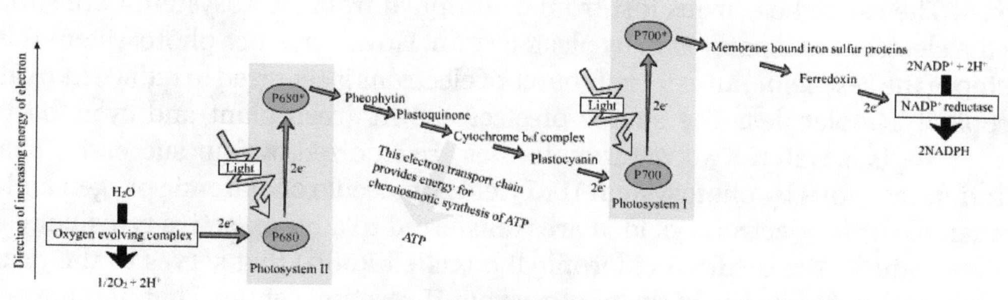

The "Z scheme"

In plants, light-dependent reactions occur in the thylakoid membranes of the chloroplasts where they drive the synthesis of ATP and NADPH. The light-dependent reactions are of two forms: cyclic and non-cyclic.

In the non-cyclic reaction, the photons are captured in the light-harvesting antenna complexes of photosystem II by chlorophyll and other accessory pigments. The absorption of a photon by the antenna complex frees an electron by a process called photoinduced charge separation. The antenna system is at the core of the chlorophyll molecule of the photosystem II reaction center. That freed electron is transferred to the primary electron-acceptor molecule, pheophytin. As the electrons are shuttled through an electron transport chain (the so-called *Z-scheme* shown in the diagram), it initially functions to generate a chemiosmotic potential by pumping proton cations (H^+) across the membrane and into the thylakoid space. An ATP synthase enzyme uses that chemiosmotic potential to make ATP during photophosphorylation, whereas NADPH is a product of the terminal redox reaction in the *Z-scheme*. The electron enters a chlorophyll molecule in Photosystem I. There it is further excited by the light absorbed by that photosystem. The electron is then passed along a chain of electron acceptors to which it transfers some of its energy. The energy delivered to the electron acceptors is used to move hydrogen ions across the thylakoid membrane into the lumen. The electron is eventually used to reduce the co-enzyme NADP with a H^+ to NADPH (which has functions in the light-independent reaction); at that point, the path of that electron ends.

The cyclic reaction is similar to that of the non-cyclic, but differs in that it generates only ATP, and no reduced NADP (NADPH) is created. The cyclic reaction takes place only at photosystem I. Once the electron is displaced from the photosystem, the electron is passed down the electron acceptor molecules and returns to photosystem I, from where it was emitted, hence the name *cyclic reaction.*

Water Photolysis

The NADPH is the main reducing agent produced by chloroplasts, which then goes on to provide a source of energetic electrons in other cellular reactions. Its production leaves chlorophyll in photosystem I with a deficit of electrons (chlorophyll has been oxidized), which must be balanced by some other reducing agent that will supply the missing electron. The excited electrons lost from chlorophyll from photosystem I are supplied from the electron transport chain by plastocyanin. However, since photosystem II is the first step of the *Z-scheme*, an external source of electrons is required to reduce its oxidized chlorophyll a molecules. The source of electrons in green-plant and cyanobacterial photosynthesis is water. Two water molecules are oxidized by four successive charge-separation reactions by photosystem II to yield a molecule of diatomic oxygen and four hydrogen ions; the electrons yielded are transferred to a redox-active tyrosine residue that then reduces the oxidized chlorophyll *a* (called P680) that serves as the primary light-driven electron donor in the photosystem II reaction center. That photo receptor is in effect reset and is then able to repeat the absorption of another photon and the release of another photo-dissociated electron. The oxidation of water is catalyzed in photosystem II by a redox-active structure that contains four manganese ions and a calcium ion; this oxygen-evolving complex binds two water molecules and contains the four oxidizing equivalents that are used to drive the water-oxidizing reaction. Photosystem II is the only known biological enzyme that carries out this oxidation of water. The hydrogen ions released contribute to the transmembrane chemiosmotic potential that leads to ATP synthesis. Oxygen is a waste product of light-dependent reactions, but the majority of organisms on Earth use oxygen for cellular respiration, including photosynthetic organisms.

Light-independent Reactions

Calvin Cycle

In the light-independent (or "dark") reactions, the enzyme RuBisCO captures CO_2 from the atmosphere and, in a process called the Calvin-Benson cycle, it uses the newly formed NADPH and releases three-carbon sugars, which are later combined to form sucrose and starch. The overall equation for the light-independent reactions in green plants is

$$3\,CO_2 + 9\,ATP + 6\,NADPH + 6\,H^+ \rightarrow C_3H_6O_3\text{-phosphate} + 9\,ADP + 8\,P_i + 6\,NADP^+ + 3\,H_2O$$

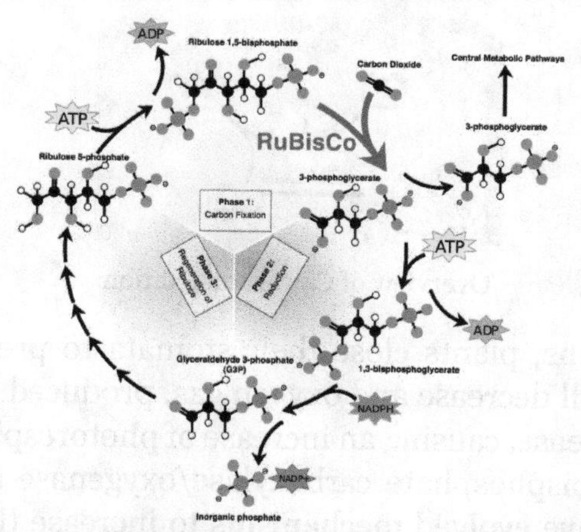

Overview of the Calvin cycle and carbon fixation

Carbon fixation produces the intermediate three-carbon sugar product, which is then converted to the final carbohydrate products. The simple carbon sugars produced by photosynthesis are then used in the forming of other organic compounds, such as the building material cellulose, the precursors for lipid and amino acid biosynthesis, or as a fuel in cellular respiration. The latter occurs not only in plants but also in animals when the energy from plants is passed through a food chain.

The fixation or reduction of carbon dioxide is a process in which carbon dioxide combines with a five-carbon sugar, ribulose 1,5-bisphosphate, to yield two molecules of a three-carbon compound, glycerate 3-phosphate, also known as 3-phosphoglycerate. Glycerate 3-phosphate, in the presence of ATP and NADPH produced during the light-dependent stages, is reduced to glyceraldehyde 3-phosphate. This product is also referred to as 3-phosphoglyceraldehyde (PGAL) or, more generically, as triose phosphate. Most (5 out of 6 molecules) of the glyceraldehyde 3-phosphate produced is used to regenerate ribulose 1,5-bisphosphate so the process can continue. The triose phosphates not thus "recycled" often condense to form hexose phosphates, which ultimately yield sucrose, starch and cellulose. The sugars produced during carbon metabolism yield carbon skeletons that can be used for other metabolic reactions like the production of amino acids and lipids.

Carbon Concentrating Mechanisms

On Land

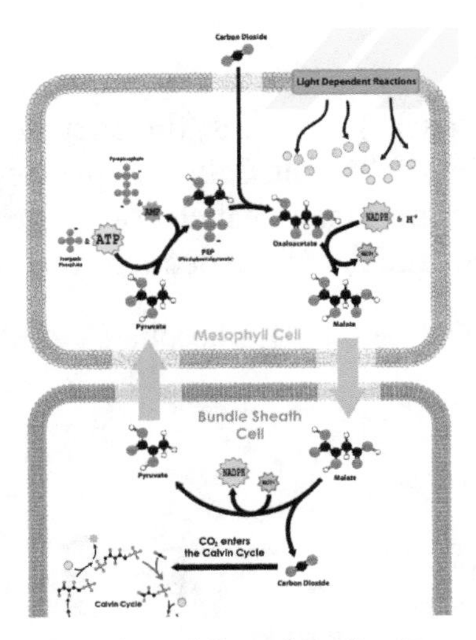

Overview of C4 carbon fixation

In hot and dry conditions, plants close their stomata to prevent water loss. Under these conditions, CO_2 will decrease and oxygen gas, produced by the light reactions of photosynthesis, will increase, causing an increase of photorespiration by the oxygenase activity of ribulose-1,5-bisphosphate carboxylase/oxygenase and decrease in carbon fixation. Some plants have evolved mechanisms to increase the CO_2 concentration in the leaves under these conditions.

Plants that use the C_4 carbon fixation process chemically fix carbon dioxide in the cells of the mesophyll by adding it to the three-carbon molecule phosphoenolpyruvate (PEP), a reaction catalyzed by an enzyme called PEP carboxylase, creating the four-carbon organic acid oxaloacetic acid. Oxaloacetic acid or malate synthesized by this process is then translocated to specialized bundle sheath cells where the enzyme RuBisCO and other Calvin cycle enzymes are located, and where CO_2 released by decarboxylation of the four-carbon acids is then fixed by RuBisCO activity to the three-carbon 3-phosphoglyceric acids. The physical separation of RuBisCO from the oxygen-generating light reactions reduces photorespiration and increases CO_2 fixation and, thus, the photosynthetic capacity of the leaf. C_4 plants can produce more sugar than C_3

plants in conditions of high light and temperature. Many important crop plants are C_4 plants, including maize, sorghum, sugarcane, and millet. Plants that do not use PEP-carboxylase in carbon fixation are called C_3 plants because the primary carboxylation reaction, catalyzed by RuBisCO, produces the three-carbon 3-phosphoglyceric acids directly in the Calvin-Benson cycle. Over 90% of plants use C_3 carbon fixation, compared to 3% that use C_4 carbon fixation; however, the evolution of C_4 in over 60 plant lineages makes it a striking example of convergent evolution.

Xerophytes, such as cacti and most succulents, also use PEP carboxylase to capture carbon dioxide in a process called Crassulacean acid metabolism (CAM). In contrast to C_4 metabolism, which *spatially* separates the CO_2 fixation to PEP from the Calvin cycle, CAM *temporally* separates these two processes. CAM plants have a different leaf anatomy from C_3 plants, and fix the CO_2 at night, when their stomata are open. CAM plants store the CO_2 mostly in the form of malic acid via carboxylation of phosphoenolpyruvate to oxaloacetate, which is then reduced to malate. Decarboxylation of malate during the day releases CO_2 inside the leaves, thus allowing carbon fixation to 3-phosphoglycerate by RuBisCO. Sixteen thousand species of plants use CAM.

In Water

Cyanobacteria possess carboxysomes, which increase the concentration of CO_2 around RuBisCO to increase the rate of photosynthesis. An enzyme, carbonic anhydrase, located within the carboxysome releases CO_2 from the dissolved hydrocarbonate ions (HCO−3). Before the CO_2 diffuses out it is quickly sponged up by RuBisCO, which is concentrated within the carboxysomes. HCO−3 ions are made from CO_2 outside the cell by another carbonic anhydrase and are actively pumped into the cell by a membrane protein. They cannot cross the membrane as they are charged, and within the cytosol they turn back into CO_2 very slowly without the help of carbonic anhydrase. This causes the HCO−3 ions to accumulate within the cell from where they diffuse into the carboxysomes. Pyrenoids in algae and hornworts also act to concentrate CO_2 around rubisco.

Order and Kinetics

The overall process of photosynthesis takes place in four stages:

Stage	Description	Time scale
1	Energy transfer in antenna chlorophyll (thylakoid membranes)	femtosecond to picosecond
2	Transfer of electrons in photochemical reactions (thylakoid membranes)	picosecond to nanosecond
3	Electron transport chain and ATP synthesis (thylakoid membranes)	microsecond to millisecond
4	Carbon fixation and export of stable products	millisecond to second

Efficiency

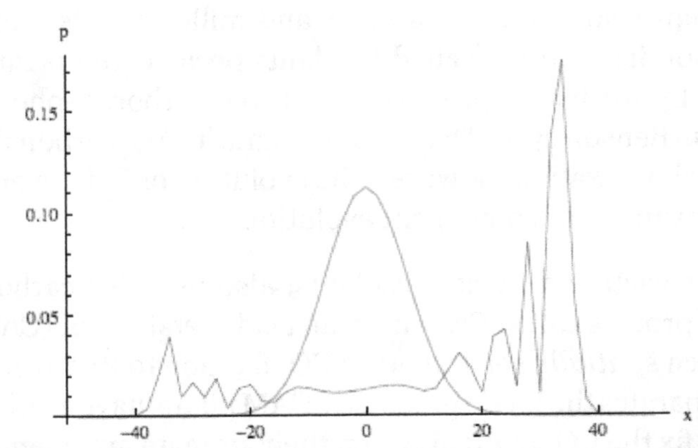

Probability distribution resulting from one-dimensional discrete time random walks. The quantum walk created using the Hadamard coin is plotted (blue) vs a classical walk (red) after 50 time steps.

Plants usually convert light into chemical energy with a photosynthetic efficiency of 3–6%. Absorbed light that is unconverted is dissipated primarily as heat, with a small fraction (1–2%) re-emitted as chlorophyll fluorescence at longer (redder) wavelengths. A fact that allows measurement of the light reaction of photosynthesis by using chlorophyll fluorometers.

Actual plants' photosynthetic efficiency varies with the frequency of the light being converted, light intensity, temperature and proportion of carbon dioxide in the atmosphere, and can vary from 0.1% to 8%. By comparison, solar panels convert light into electric energy at an efficiency of approximately 6–20% for mass-produced panels, and above 40% in laboratory devices.

The efficiency of both light and dark reactions can be measured but the relationship between the two can be complex. For example, the ATP and NADPH energy molecules, created by the light reaction, can be used for carbon fixation or for photorespiration in C_3 plants. Electrons may also flow to other electron sinks. For this reason, it is not uncommon for authors to differentiate between work done under non-photorespiratory conditions and under photorespiratory conditions.

Chlorophyll fluorescence of photosystem II can measure the light reaction, and Infrared gas analyzers can measure the dark reaction. It is also possible to investigate both at the same time using an integrated chlorophyll fluorometer and gas exchange system, or by using two separate systems together. Infrared gas analyzers and some moisture sensors are sensitive enough to measure the photosynthetic assimilation of CO_2, and of ΔH_2O using reliable methods CO_2 is commonly measured in μmols/m²/s⁻¹, parts per million or volume per million and H_2O is commonly measured in mmol/m²/s⁻¹ or in mbars. By measuring CO_2 assimilation, ΔH_2O, leaf temperature, barometric pressure, leaf area, and photosynthetically active radiation or PAR, it becomes possible to estimate, "A" or carbon assimilation, "E" or transpiration, "gs" or stomatal conductance, and

Ci or intracellular CO_2. However, it is more common to used chlorophyll fluorescence for plant stress measurement, where appropriate, because the most commonly used measuring parameters FV/FM and Y(II) or F/FM' can be made in a few seconds, allowing the measurement of larger plant populations.

Gas exchange systems that offer control of CO_2 levels, above and below ambient, allow the common practice of measurement of A/Ci curves, at different CO_2 levels, to characterize a plant's photosynthetic response.

Integrated chlorophyll fluorometer – gas exchange systems allow a more precise measure of photosynthetic response and mechanisms. While standard gas exchange photosynthesis systems can measure Ci, or substomatal CO_2 levels, the addition of integrated chlorophyll fluorescence measurements allows a more precise measurement of C_C to replace Ci. The estimation of CO_2 at the site of carboxylation in the chloroplast, or C_C, becomes possible with the measurement of mesophyll conductance or g_m using an integrated system.

Photosynthesis measurement systems are not designed to directly measure the amount of light absorbed by the leaf. But analysis of chlorophyll-fluorescence, P700- and P515-absorbance and gas exchange measurements reveal detailed information about e.g. the photosystems, quantum efficiency and the CO_2 assimilation rates. With some instruments even wavelength-dependency of the photosynthetic efficiency can be analyzed.

A phenomenon known as quantum walk increases the efficiency of the energy transport of light significantly. In the photosynthetic cell of an algae, bacterium, or plant, there are light-sensitive molecules called chromophores arranged in an antenna-shaped structure named a photocomplex. When a photon is absorbed by a chromophore, it is converted into a quasiparticle referred to as an exciton, which jumps from chromophore to chromophore towards the reaction center of the photocomplex, a collection of molecules that traps its energy in a chemical form that makes it accessible for the cell's metabolism. The exciton's wave properties enable it to cover a wider area and try out several possible paths simultaneously, allowing it to instantaneously "choose" the most efficient route, where it will have the highest probability of arriving at its destination in the minimum possible time. Because that quantum walking takes place at temperatures far higher than quantum phenomena usually occur, it is only possible over very short distances, due to obstacles in the form of destructive interference that come into play. These obstacles cause the particle to lose its wave properties for an instant before it regains them once again after it is freed from its locked position through a classic "hop". The movement of the electron towards the photo center is therefore covered in a series of conventional hops and quantum walks.

Evolution

Early photosynthetic systems, such as those in green and purple sulfur and green and purple nonsulfur bacteria, are thought to have been anoxygenic, and used various other molecules as electron donors rather than water. Green and purple sulfur bacteria are thought to have used hydrogen and sulfur as electron donors. Green nonsulfur bacteria used various amino and other organic acids as an electron donor. Purple nonsulfur bacteria used a variety of nonspecific organic molecules. The use of these molecules is consistent with the geological evidence that Earth's early atmosphere was highly reducing at that time.

Fossils of what are thought to be filamentous photosynthetic organisms have been dated at 3.4 billion years old.

The main source of oxygen in the Earth's atmosphere derives from oxygenic photosynthesis, and its first appearance is sometimes referred to as the oxygen catastrophe. Geological evidence suggests that oxygenic photosynthesis, such as that in cyanobacteria, became important during the Paleoproterozoic era around 2 billion years ago. Modern photosynthesis in plants and most photosynthetic prokaryotes is oxygenic. Oxygenic photosynthesis uses water as an electron donor, which is oxidized to molecular oxygen (O2) in the photosynthetic reaction center.

Symbiosis and the Origin of Chloroplasts

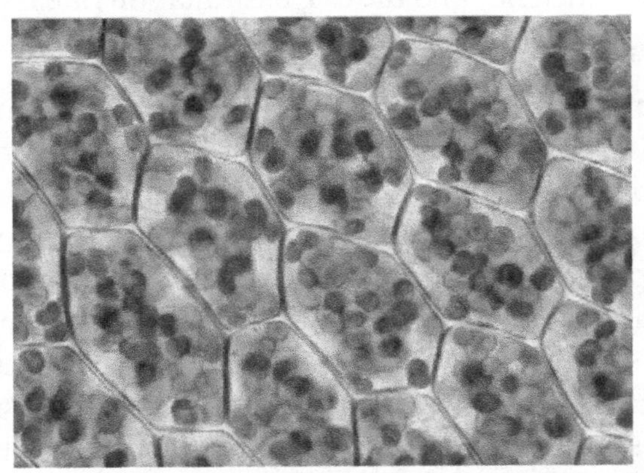

Plant cells with visible chloroplasts (from a moss, *Plagiomnium affine*)

Several groups of animals have formed symbiotic relationships with photosynthetic algae. These are most common in corals, sponges and sea anemones. It is presumed that this is due to the particularly simple body plans and large surface areas of these animals compared to their volumes. In addition, a few marine mollusks *Elysia viridis* and *Elysia chlorotica* also maintain a symbiotic relationship with chloroplasts they capture from the algae in their diet and then store in their bodies. This allows the mollusks to survive solely by photosynthesis for several months at a time. Some of the genes from the plant cell nucleus have even been transferred to the slugs, so that the chloroplasts can be supplied with proteins that they need to survive.

An even closer form of symbiosis may explain the origin of chloroplasts. Chloroplasts have many similarities with photosynthetic bacteria, including a circular chromosome, prokaryotic-type ribosome, and similar proteins in the photosynthetic reaction center. The endosymbiotic theory suggests that photosynthetic bacteria were acquired (by endocytosis) by early eukaryotic cells to form the first plant cells. Therefore, chloroplasts may be photosynthetic bacteria that adapted to life inside plant cells. Like mitochondria, chloroplasts possess their own DNA, separate from the nuclear DNA of their plant host cells and the genes in this chloroplast DNA resemble those found in cyanobacteria. DNA in chloroplasts codes for redox proteins such as those found in the photosynthetic reaction centers. The CoRR Hypothesis proposes that this **Co**-location is required for Redox Regulation.

Cyanobacteria and the Evolution of Photosynthesis

The biochemical capacity to use water as the source for electrons in photosynthesis evolved once, in a common ancestor of extant cyanobacteria. The geological record indicates that this transforming event took place early in Earth's history, at least 2450–2320 million years ago (Ma), and, it is speculated, much earlier. Because the Earth's atmosphere contained almost no oxygen during the estimated development of photosynthesis, it is believed that the first photosynthetic cyanobacteria did not generate oxygen. Available evidence from geobiological studies of Archean (>2500 Ma) sedimentary rocks indicates that life existed 3500 Ma, but the question of when oxygenic photosynthesis evolved is still unanswered. A clear paleontological window on cyanobacterial evolution opened about 2000 Ma, revealing an already-diverse biota of blue-green algae. Cyanobacteria remained the principal primary producers of oxygen throughout the Proterozoic Eon (2500–543 Ma), in part because the redox structure of the oceans favored photoautotrophs capable of nitrogen fixation. Green algae joined blue-green algae as the major primary producers of oxygen on continental shelves near the end of the Proterozoic, but it was only with the Mesozoic (251–65 Ma) radiations of dinoflagellates, coccolithophorids, and diatoms did the primary production of oxygen in marine shelf waters take modern form. Cyanobacteria remain critical to marine ecosystems as primary producers of oxygen in oceanic gyres, as agents of biological nitrogen fixation, and, in modified form, as the plastids of marine algae.

Discovery

Although some of the steps in photosynthesis are still not completely understood, the overall photosynthetic equation has been known since the 19th century.

Jan van Helmont began the research of the process in the mid-17th century when he carefully measured the mass of the soil used by a plant and the mass of the plant as it grew. After noticing that the soil mass changed very little, he hypothesized that the mass of the growing plant must come from the water, the only substance he added to the potted plant. His hypothesis was partially accurate — much of the gained mass also comes from carbon dioxide as well as water. However, this was a signaling point to the idea that the bulk of a plant's biomass comes from the inputs of photosynthesis, not the soil itself.

Joseph Priestley, a chemist and minister, discovered that, when he isolated a volume of air under an inverted jar, and burned a candle in it, the candle would burn out very quickly, much before it ran out of wax. He further discovered that a mouse could similarly "injure" air. He then showed that the air that had been "injured" by the candle and the mouse could be restored by a plant.

In 1778, Jan Ingenhousz, repeated Priestley's experiments. He discovered that it was the influence of sunlight on the plant that could cause it to revive a mouse in a matter of hours.

In 1796, Jean Senebier, a Swiss pastor, botanist, and naturalist, demonstrated that green plants consume carbon dioxide and release oxygen under the influence of light. Soon afterward, Nicolas-Théodore de Saussure showed that the increase in mass of the plant as it grows could not be due only to uptake of CO_2 but also to the incorporation of water. Thus, the basic reaction by which photosynthesis is used to produce food (such as glucose) was outlined.

Cornelis Van Niel made key discoveries explaining the chemistry of photosynthesis. By studying purple sulfur bacteria and green bacteria he was the first to demonstrate that photosynthesis is a light-dependent redox reaction, in which hydrogen reduces carbon dioxide.

Robert Emerson discovered two light reactions by testing plant productivity using different wavelengths of light. With the red alone, the light reactions were suppressed. When blue and red were combined, the output was much more substantial. Thus, there were two photosystems, one absorbing up to 600 nm wavelengths, the other up to 700 nm. The former is known as PSII, the latter is PSI. PSI contains only chlorophyll "a", PSII contains primarily chlorophyll "a" with most of the available chlorophyll "b", among other pigment. These include phycobilins, which are the red and blue pigments of red and blue algae respectively, and fucoxanthol for brown algae and diatoms. The process is most productive when the absorption of quanta are equal in both the PSII and PSI, assuring that input energy from the antenna complex is divided between the PSI and PSII system, which in turn powers the photochemistry.

Melvin Calvin works in his photosynthesis laboratory.

Robert Hill thought that a complex of reactions consisting of an intermediate to cytochrome b_6 (now a plastoquinone), another is from cytochrome f to a step in the carbohydrate-generating mechanisms. These are linked by plastoquinone, which does require energy to reduce cytochrome f for it is a sufficient reductant. Further experiments to prove that the oxygen developed during the photosynthesis of green plants came from water, were performed by Hill in 1937 and 1939. He showed that isolated chloroplasts give off oxygen in the presence of unnatural reducing agents like iron oxalate, ferricyanide or benzoquinone after exposure to light. The Hill reaction is as follows:

$$2\ H_2O + 2\ A + (\text{light, chloroplasts}) \rightarrow 2\ AH_2 + O_2$$

where A is the electron acceptor. Therefore, in light, the electron acceptor is reduced and oxygen is evolved.

Samuel Ruben and Martin Kamen used radioactive isotopes to determine that the oxygen liberated in photosynthesis came from the water.

Melvin Calvin and Andrew Benson, along with James Bassham, elucidated the path of carbon assimilation (the photosynthetic carbon reduction cycle) in plants. The carbon reduction cycle is known as the Calvin cycle, which ignores the contribution of Bassham and Benson. Many scientists refer to the cycle as the Calvin-Benson Cycle, Benson-Calvin, and some even call it the Calvin-Benson-Bassham (or CBB) Cycle.

Nobel Prize-winning scientist Rudolph A. Marcus was able to discover the function and significance of the electron transport chain.

Otto Heinrich Warburg and Dean Burk discovered the I-quantum photosynthesis reaction that splits the CO_2, activated by the respiration.

Louis N.M. Duysens and Jan Amesz discovered that chlorophyll a will absorb one light, oxidize cytochrome f, chlorophyll a (and other pigments) will absorb another light, but will reduce this same oxidized cytochrome, stating the two light reactions are in series.

Development of the Concept

In 1893, Charles Reid Barnes proposed two terms, *photosyntax* and *photosynthesis*, for the biological process of *synthesis of complex carbon compounds out of carbonic acid, in the presence of chlorophyll, under the influence of light*. Over time, the term *photosynthesis* came into common usage as the term of choice. Later discovery of anoxygenic photosynthetic bacteria and photophosphorylation necessitated redefinition of the term.

C3 : C4 Photosynthesis Research

After WWII at late 1940 at the University of California, Berkeley, the details of photosynthetic carbon metabolism were sorted out by the chemists Melvin Calvin,

Andrew Benson, James Bassham and a score of students and researchers utilizing the carbon-14 isotope and paper chromatography techniques. The pathway of CO2 fixation by the algae *Chlorella* in a fraction of a second in light resulted in a 3 carbon molecule called phosphoglyceric acid (PGA). For that original and ground-breaking work, a Nobel Prize in Chemistry was awarded to Melvin Calvin 1961. In parallel, plant physiologists studied leaf gas exchanges using the new method of infrared gas analysis and a leaf chamber where the net photosynthetic rates ranged from 10 to 13 u mole CO2/square metere.sec., with the conclusion that all terrestrial plants having the same photosynthetic capacities that were light saturated at less than 50% of sunlight. These rates were determined in potted plants grown indoors under low light intensity.

Later in 1958-1963 at Cornell University, field grown maize was reported to have much greater leaf photosynthetic rates of 40 u mol CO2/square meter.sec and was not saturated at near full sunlight. This higher rate in maize was almost double those observed in other species such as wheat and soybean, indicating that large differences in photosynthesis exist among higher plants. At the University of Arizona, detailed gas exchange research on more than 15 species of monocot and dicot uncovered for the first time that differences in leaf anatomy are crucial factors in differentiating photosynthetic capacities among species. In tropical grasses, including maize, sorghum, sugarcane, Bermuda grass and in the dicot amaranthus, leaf photosynthetic rates were around 38–40 u mol CO2/square meter.sec., and the leaves have two types of green cells, i. e. outer layer of mesophyll cells surrounding a tightly packed cholorophyllous vascular bundle sheath cells. This type of anatomy was termed Kranz anatomy in the 19th century by the botanist Gottlieb Haberlandt while studying leaf anatomy of sugarcane. Plant species with the greatest photosynthetic rates and Kranz anatomy showed no apparent photorespiration, very low CO2 compensation point, high optimum temperature, high stomatal resistances and lower mesophyll resistances for gas diffusion and rates never saturated at full sun light. The research at Arizona was designated Citation Classic by the ISI 1986. These species was later termed C4 plants as the first stable compound of CO2 fixation in light has 4 carbon as malate and aspartate. Other species that lack Kranz anatomy were termed C3 type such as cotton and sunflower, as the first stable carbon compound is the 3-carbon PGA acid. At 1000 ppm CO2 in measuring air, both the C3 and C4 plants had similar leaf photosynthetic rates around 60 u mole CO2/ square meter.sec. indicating the suppression of phototorespiration in C3 plants.

Factors

There are three main factors affecting photosynthesis and several corollary factors. The three main are:

- Light irradiance and wavelength
- Carbon dioxide concentration
- Temperature.

The leaf is the primary site of photosynthesis in plants.

Light Intensity (Irradiance), Wavelength and Temperature

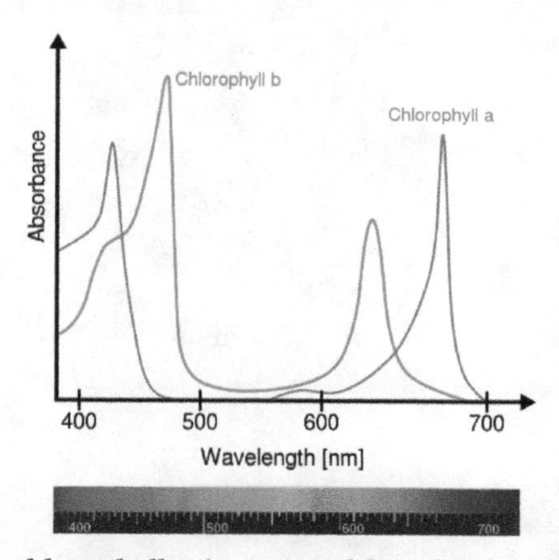

Absorbance spectra of free chlorophyll *a* (green) and *b* (red) in a solvent. The action spectra of chlorophyll molecules are slightly modified *in vivo* depending on specific pigment-protein interactions.

The process of photosynthesis provides the main input of free energy into the biosphere, and is one of four main ways in which radiation is important for plant life.

The radiation climate within plant communities is extremely variable, with both time and space.

In the early 20th century, Frederick Blackman and Gabrielle Matthaei investigated the effects of light intensity (irradiance) and temperature on the rate of carbon assimilation.

- At constant temperature, the rate of carbon assimilation varies with irradiance, increasing as the irradiance increases, but reaching a plateau at higher irradiance.
- At low irradiance, increasing the temperature has little influence on the rate of carbon assimilation. At constant high irradiance, the rate of carbon assimilation increases as the temperature is increased.

These two experiments illustrate several important points: First, it is known that, in general, photochemical reactions are not affected by temperature. However, these experiments clearly show that temperature affects the rate of carbon assimilation, so there must be two sets of reactions in the full process of carbon assimilation. These are, of course, the light-dependent 'photochemical' temperature-independent stage, and the light-independent, temperature-dependent stage. Second, Blackman's experiments illustrate the concept of limiting factors. Another limiting factor is the wavelength of light. Cyanobacteria, which reside several meters underwater, cannot receive the correct wavelengths required to cause photoinduced charge separation in conventional photosynthetic pigments. To combat this problem, a series of proteins with different pigments surround the reaction center. This unit is called a phycobilisome.

Carbon Dioxide Levels and Photorespiration

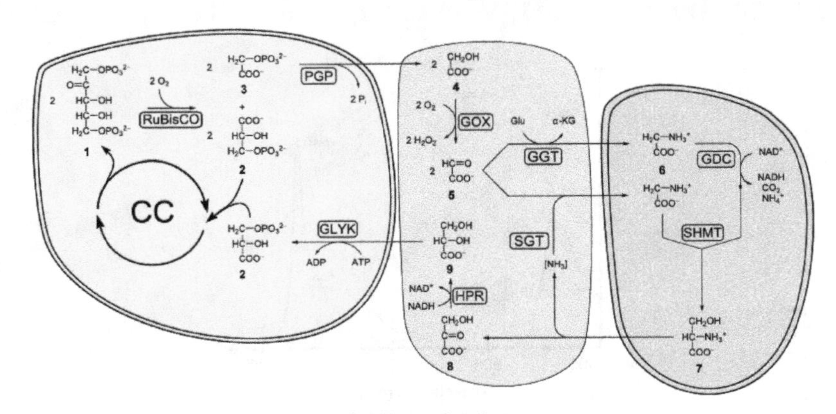

Photorespiration

As carbon dioxide concentrations rise, the rate at which sugars are made by the light-independent reactions increases until limited by other factors. RuBisCO, the enzyme that captures carbon dioxide in the light-independent reactions, has a binding affinity for both carbon dioxide and oxygen. When the concentration of carbon dioxide is high, RuBisCO will fix carbon dioxide. However, if the carbon dioxide concentration is low, RuBisCO will bind oxygen instead of carbon dioxide. This process, called photorespiration, uses energy, but does not produce sugars.

RuBisCO oxygenase activity is disadvantageous to plants for several reasons:

1. One product of oxygenase activity is phosphoglycolate (2 carbon) instead of 3-phosphoglycerate (3 carbon). Phosphoglycolate cannot be metabolized by the Calvin-Benson cycle and represents carbon lost from the cycle. A high oxygenase activity, therefore, drains the sugars that are required to recycle ribulose 5-bisphosphate and for the continuation of the Calvin-Benson cycle.

2. Phosphoglycolate is quickly metabolized to glycolate that is toxic to a plant at a high concentration; it inhibits photosynthesis.

3. Salvaging glycolate is an energetically expensive process that uses the glycolate pathway, and only 75% of the carbon is returned to the Calvin-Benson cycle as 3-phosphoglycerate. The reactions also produce ammonia (NH_3), which is able to diffuse out of the plant, leading to a loss of nitrogen.

   A highly simplified summary is:

   2 glycolate + ATP → 3-phosphoglycerate + carbon dioxide + ADP + NH_3

The salvaging pathway for the products of RuBisCO oxygenase activity is more commonly known as photorespiration, since it is characterized by light-dependent oxygen consumption and the release of carbon dioxide.

Light-dependent Reactions

In photosynthesis, the light-dependent reactions take place on the thylakoid membranes. The inside of the thylakoid membrane is called the lumen, and outside the thylakoid membrane is the stroma, where the light-independent reactions take place. The thylakoid membrane contains some integral membrane protein complexes that catalyze the light reactions. There are four major protein complexes in the thylakoid membrane: Photosystem II (PSII), Cytochrome b6f complex, Photosystem I (PSI), and ATP synthase. These four complexes work together to ultimately create the products ATP and NADPH.

The two photosystems absorb light energy through pigments - primarily the chlorophylls, which are responsible for the green color of leaves. The light-dependent reactions begin in photosystem II. When a chlorophyll *a* molecule within the reaction center of PSII absorbs a photon, an electron in this molecule attains a higher energy level. Because this state of an electron is very unstable, the electron is transferred from one to another molecule creating a chain of redox reactions, called an electron transport chain (ETC). The electron flow goes from PSII to cytochrome b6f to PSI. In PSI, the electron gets the energy from another photon. The final electron acceptor is NADP. In oxygenic photosynthesis, the first electron donor is water, creating oxygen as a waste product. In anoxygenic photosynthesis various electron donors are used.

Cytochrome b6f and ATP synthase work together to create ATP. This process is called photophosphorylation, which occurs in two different ways. In non-cyclic photophosphorylation, cytochrome b6f uses the energy of electrons from PSII to pump protons from the stroma to the lumen. The proton gradient across the thylakoid membrane creates a proton-motive force, used by ATP synthase to form ATP. In cyclic photophosphorylation, cytochrome b6f uses the energy of electrons from not only PSII but also PSI to create more ATP and to stop the production of NADPH. Cyclic phosphorylation is important to create ATP and maintain NADPH in the right proportion for the light-independent reactions.

The net-reaction of all light-dependent reactions in oxygenic photosynthesis is:

$$2H2O + 2NADP++ 3ADP + 3P_i \rightarrow O2 + 2NADPH + 3ATP$$

The two photosystems are protein complexes that absorb photons and are able to use this energy to create an electron transport chain. Photosystem I and II are very similar in structure and function. They use special proteins, called light-harvesting complexes, to absorb the photons with very high effectiveness. If a special pigment molecule in a photosynthetic reaction center absorbs a photon, an electron in this pigment attains the excited state and then is transferred to another molecule in the reaction center. This reaction, called photoinduced charge separation, is the start of the electron flow and is unique because it transforms light energy into chemical forms.

The Reaction Center

The reaction center is in the thylakoid membrane. It transfers light energy to a dimer of chlorophyll pigment molecules near the periplasmic (or thylakoid lumen) side of the membrane. This dimer is called a special pair because of its fundamental role in photosynthesis. This special pair is slightly different in PSI and PSII reaction center. In PSII, it absorbs photons with a wavelength of 680 nm, and it is therefore called P680. In PSI, it absorbs photons at 700 nm, and it is called P700. In bacteria, the special pair is called P760, P840, P870, or P960.

If an electron of the special pair in the reaction center becomes excited, it cannot transfer this energy to another pigment using resonance energy transfer. In normal circumstances, the electron should return to the ground state, but, because the reaction center is arranged so that a suitable electron acceptor is nearby, the excited electron can move from the initial molecule to the acceptor. This process results in the formation of a positive charge on the special pair (due to the loss of an electron) and a negative charge on the acceptor and is, hence, referred to as photoinduced charge separation. In other words, electrons in pigment molecules can exist at specific energy levels. Under normal circumstances, they exist at the lowest possible energy level they can. However, if there is enough energy to move them into the next energy level, they can absorb that energy and occupy that higher energy level. The light they absorb contains the necessary amount of energy needed to push them into the next level. Any light that does not have enough or has too much energy cannot be absorbed and is reflected. The electron in the higher energy level, however, does not want to be there; the electron is unstable and must return to its normal lower energy level. To do this, it must release the energy that has put it into the higher energy state to begin with. This can happen various ways. The extra energy can be converted into molecular motion and lost as heat. Some of the extra energy can be lost as heat energy, while the rest is lost as light. This re-emission of light energy is called fluorescence. The energy, but not the e- itself, can be passed onto another molecule. This is called resonance. The energy and the e- can be transferred to another molecule. Plant pigments usually utilize the last two of these reactions to convert the sun's energy into their own.

This initial charge separation occurs in less than 10 picoseconds (10^{-11} seconds). In their high-energy states, the special pigment and the acceptor could undergo charge recombination; that is, the electron on the acceptor could move back to neutralize the positive charge on the special pair. Its return to the special pair would waste a valuable high-energy electron and simply convert the absorbed light energy into heat. In the case of PSII, this backflow of electrons can produce reactive oxygen species leading to photoinhibition. Three factors in the structure of the reaction center work together to suppress charge recombination nearly completely.

- Another electron acceptor is less than 10 Å away from the first acceptor, and so the electron is rapidly transferred farther away from the special pair.
- An electron donor is less than 10 Å away from the special pair, and so the positive charge is neutralized by the transfer of another electron
- The electron transfer back from the electron acceptor to the positively charged special pair is especially slow. The rate of an of electron transfer reaction increases with its thermodynamic favorability up to a point and then decreases. The back transfer is so favourable that it takes place in the inverted region where electron-transfer rates become slower.

Thus, electron transfer proceeds efficiently from the first electron acceptor to the next, creating an electron transport chain that ends if it has reached NADPH.

Photosynthetic Electron Transport Chains in Chloroplasts

The photosynthesis process in chloroplasts begins when an electron of P680 of PSII attains a higher-energy level. This energy is used to reduce a chain of electron acceptors that have subsequently lowered redox-potentials. This chain of electron acceptors is known as an electron transport chain. When this chain reaches PS I, an electron is again excited, creating a high redox-potential. The electron transport chain of photosynthesis is often put in a diagram called the z-scheme, because the redox diagram from P680 to P700 resembles the letter z.

The final product of PSII is plastoquinol, a mobile electron carrier in the membrane. Plastoquinol transfers the electron from PSII to the proton pump, cytochrome b6f. The ultimate electron donor of PSII is water. Cytochrome b6f proceeds the electron chain to PSI through plastocyanin molecules. PSI is able to continue the electron transfer in two different ways. It can transfer the electrons either to plastoquinol again, creating a cyclic electron flow, or to an enzyme called FNR (Ferredoxin—NADP(+) reductase), creating a non-cyclic electron flow. PSI releases FNR into the stroma, where it reduces NADP+to NADPH.

Activities of the electron transport chain, especially from cytochrome b6f, lead to pumping of protons from the stroma to the lumen. The resulting transmembrane proton gradient is used to make ATP via ATP synthase.

The overall process of the photosynthetic electron transport chain in chloroplasts is:

$$H2O \rightarrow PS\ II \rightarrow plastoquinone \rightarrow cyt_6^b \rightarrow plastocyanin \rightarrow PS\ I \rightarrow NADPH$$

Photosystem II

PS II is an extremely complex, highly organized transmembrane structure that contains a *water-splitting complex*, chlorophylls and carotenoid pigments, a *reaction center* (P680), pheophytin (a pigment similar to chlorophyll), and two quinones. It uses the energy of sunlight to transfer electrons from water to a mobile electron carrier in the membrane called *plastoquinone*:

$$H2O \rightarrow P680 \rightarrow P680^* \rightarrow plastoquinone$$

Plastoquinone, in turn, transfers electrons to cyt_6^b, which feeds them into PS I.

The Water-splitting Complex

The step H2O → P680 is performed by a poorly understood structure embedded within PS II called the water-splitting complex or the oxygen-evolving complex. It catalyzes a reaction that splits water into electrons, protons and oxygen:

$$2H2O \rightarrow 4H^+ + 4e^- + O2$$

The electrons are transferred to special chlorophyll molecules (embedded in PS II) that are promoted to a higher-energy state by the energy of photons.

The Reaction Center

The excitation P680 → P680*of the reaction center pigment P680 occurs here. These special chlorophyll molecules embedded in PS II absorb the energy of photons, with maximal absorption at 680 nm. Electrons within these molecules are promoted to a higher-energy state. This is one of two core processes in photosynthesis, and it occurs with astonishing efficiency (greater than 90%) because, in addition to direct excitation by light at 680 nm, the energy of light first harvested by antenna proteins at other wavelengths in the light-harvesting system is also transferred to these special chlorophyll molecules.

This is followed by the step P680*→ pheophytin, and then on to plastoquinone, which occurs within the reaction center of PS II. High-energy electrons are transferred to plastoquinone before it subsequently picks up two protons to become plastoquinol. Plastoquinol is then released into the membrane as a mobile electron carrier.

This is the second core process in photosynthesis. The initial stages occur within *picoseconds*, with an efficiency of 100%. The seemingly impossible efficiency is due to the precise positioning of molecules within the reaction center. This is a solid-state process, not a chemical reaction. It occurs within an essentially crystalline environment created by the macromolecular structure of PS II. The usual rules of chemistry (which involve random collisions and random energy distributions) do not apply in solid-state environments.

Link of Water-splitting Complex and Chlorophyll Excitation

When the chlorophyll passes the electron to pheophytin, it obtains an electron from P_{680}^{*}. In turn, P_{680}^{*} can oxidize the Z (or Y_Z) molecule. Once oxidized, the Z molecule can derive electrons from the oxygen-evolving complex.

Summary

PS II is a transmembrane structure found in all chloroplasts. It splits water into electrons, protons and molecular oxygen. The electrons are transferred to plastoquinone, which carries them to a proton pump. Molecular oxygen is released into the atmosphere.

The emergence of such an incredibly complex structure, a macromolecule that converts the energy of sunlight into potentially useful work with efficiencies that are impossible in ordinary experience, seems almost magical at first glance. Thus, it is of considerable interest that, in essence, the same structure is found in purple bacteria.

Cytochrome $_6^b$

PS II and PS I are connected by a transmembrane proton pump, cytochrome *b6* complex (plastoquinol—plastocyanin reductase; EC 1.10.99.1). Electrons from PS II are carried by plastoquinol to cyt$_6^b$, where they are removed in a stepwise fashion (reforming plastoquinone) and transferred to a water-soluble electron carrier called *plastocyanin*. This redox process is coupled to the pumping of four protons across the membrane. The resulting proton gradient (together with the proton gradient produced by the water-splitting complex in PS II) is used to make ATP via ATP synthase.

The similarity in structure and function between cytochrome $_6^b$ (in chloroplasts) and cytochrome *bc1* (*Complex III* in mitochondria) is striking. Both are transmembrane structures that remove electrons from a mobile, lipid-soluble electron carrier (plastoquinone in chloroplasts; ubiquinone in mitochondria) and transfer them to a mobile, water-soluble electron carrier (plastocyanin in chloroplasts; cytochrome *c* in mitochondria). Both are proton pumps that produce a transmembrane proton gradient.

Photosystem I

The cyclic light-dependent reactions occur when only the sole photosystem being used is photosystem 1. Photosystem 1 excites electrons which then cycle from the transport protein, ferredoxin (Fd), to the cytochrome complex, b6f, to another transport protein, plastocyanin (Pc), and back to photosystem I. A proton gradient is created across the thylakoid membrane (6) as protons (3) are transported from the chloroplast stroma (4) to the thylakoid lumen (5). Through chemiosmosis, ATP (9) is produced where ATP synthase (1) binds an inorganic phosphate group (8) to an ADP molecule (7).

PS I accepts electrons from plastocyanin and transfers them either to NADPH (*noncyclic electron transport*) or back to cytochrome $^{b}_{6}$ (*cyclic electron transport*):

plastocyanin → P700 → P700* → FNR → NADPH

↑ ↓

$^{b}_{6}$ ← plastoquinone

PS I, like PS II, is a complex, highly organized transmembrane structure that contains antenna chlorophylls, a reaction center (P700), phylloquinine, and a number of iron-sulfur proteins that serve as intermediate redox carriers.

The light-harvesting system of PS I uses multiple copies of the same transmembrane proteins used by PS II. The energy of absorbed light (in the form of delocalized, high-energy electrons) is funneled into the reaction center, where it excites special chlorophyll molecules (P700, maximum light absorption at 700 nm) to a higher energy level. The process occurs with astonishingly high efficiency.

Electrons are removed from excited chlorophyll molecules and transferred through a series of intermediate carriers to *ferredoxin*, a water-soluble electron carrier. As in PS II, this is a solid-state process that operates with 100% efficiency.

There are two different pathways of electron transport in PS I. In *noncyclic electron transport*, ferredoxin carries the electron to the enzyme ferredoxin NADP+oxidoreductase (FNR) that reduces NADP+to NADPH. In *cyclic electron transport*, electrons from ferredoxin are transferred (via plastoquinone) to a proton pump, cytochrome$^{b}_{6}$. They are then returned (via plastocyanin) to P700.

NADPH and ATP are used to synthesize organic molecules from CO2. The ratio of NADPH to ATP production can be adjusted by adjusting the balance between cyclic and noncyclic electron transport.

It is noteworthy that PS I closely resembles photosynthetic structures found in green sulfur bacteria, just as PS II resembles structures found in purple bacteria.

Photosynthetic Electron Transport Chains in Bacteria

PS II, PS I, and cytochrome$^{b}_{6}$ are found in chloroplasts. All plants and all photosynthetic algae contain chloroplasts, which produce NADPH and ATP by the mechanisms described above. In essence, the same transmembrane structures are also found in *cyanobacteria.*

Unlike plants and algae, cyanobacteria are prokaryotes. They do not contain chloroplasts. Rather, they bear a striking resemblance to chloroplasts themselves. This suggests that organisms resembling cyanobacteria were the evolutionary precursors of chloroplasts. One imagines primitive eukaryotic cells taking up cyanobacteria as intracellular symbionts in a process known as endosymbiosis.

Cyanobacteria

Cyanobacteria contain structures similar to PS II and PS I in chloroplasts. Their light-harvesting system is different from that found in plants (they use *phycobilins*, rather than chlorophylls, as antenna pigments), but their electron transport chain

$$H_2O \rightarrow \text{PS II} \rightarrow \text{plastoquinone} \rightarrow b_6 \rightarrow \text{cytochrome } c_6 \rightarrow \text{PS I} \rightarrow \text{ferredoxin} \rightarrow \text{NADPH}$$

$$\uparrow \qquad\qquad \downarrow$$

$$b_6 \leftarrow \text{plastoquinone}$$

is, in essence, the same as the electron transport chain in chloroplasts. The mobile water-soluble electron carrier is cytochrome c_6 in cyanobacteria, plastocyanin in plants.

Cyanobacteria can also synthesize ATP by oxidative phosphorylation, in the manner of other bacteria. The electron transport chain is

$$\text{NADH dehydrogenase} \rightarrow \text{plastoquinone} \rightarrow b_6 \rightarrow \text{cytochrome } c_6 \rightarrow \text{cytochrome } aa_3 \rightarrow O_2$$

where the mobile electron carriers are plastoquinone and cytochrome c_6, while the proton pumps are NADH dehydrogenase, b_6 and cytochrome aa_3.

Cyanobacteria are the only bacteria that produce oxygen during photosynthesis. Earth's primordial atmosphere was anoxic. Organisms like cyanobacteria produced our present-day oxygen-containing atmosphere.

The other two major groups of photosynthetic bacteria, purple bacteria and green sulfur bacteria, contain only a single photosystem and do not produce oxygen.

Purple Bacteria

Purple bacteria contain a single photosystem that is structurally related to PS II in cyanobacteria and chloroplasts:

$$P870 \rightarrow P870^* \rightarrow \text{ubiquinone} \rightarrow bc_1 \rightarrow \text{cytochrome } c_2 \rightarrow P870$$

This is a *cyclic* process in which electrons are removed from an excited chlorophyll molecule (*bacteriochlorophyll*; P870), passed through an electron transport chain to a proton pump (cytochrome bc_1 complex, similar but not identical to cytochrome bc_1 in chloroplasts), and then returned to the chlorophyll molecule. The result is a proton gradient, which is used to make ATP via ATP synthase. As in cyanobacteria and chloroplasts, this is a solid-state process that depends on the precise orientation of various functional groups within a complex transmembrane macromolecular structure.

To make NADPH, purple bacteria use an external electron donor (hydrogen, hydrogen sulfide, sulfur, sulfite, or organic molecules such as succinate and lactate) to feed electrons into a reverse electron transport chain.

Green Sulfur Bacteria

Green sulfur bacteria contain a photosystem that is analogous to PS I in chloroplasts:

P840 → P840* → ferredoxin → NADH

↑ ↓

cyt c_{553} ← bc_1 ← menaquinone

There are two pathways of electron transfer. In *cyclic electron transfer*, electrons are removed from an excited chlorophyll molecule, passed through an electron transport chain to a proton pump, and then returned to the chlorophyll. The mobile electron carriers are, as usual, a lipid-soluble quinone and a water-soluble cytochrome. The resulting proton gradient is used to make ATP.

In *noncyclic electron transfer*, electrons are removed from an excited chlorophyll molecule and used to reduce NAD^+ to NADH. The electrons removed from P840 must be replaced. This is accomplished by removing electrons from H2S, which is oxidized to sulfur (hence the name "green *sulfur* bacteria").

Purple bacteria and green sulfur bacteria occupy relatively minor ecological niches in the present day biosphere. They are of interest because of their importance in precambrian ecologies, and because their methods of photosynthesis were the likely evolutionary precursors of those in modern plants.

History

The first ideas about light being used in photosynthesis were proposed by Colin Flannery in 1779 who recognized it was sunlight falling on plants that was required, although Joseph Priestley had noted the production of oxygen without the association with light in 1772. Cornelius Van Niel proposed in 1931 that photosynthesis is a case of general mechanism where a photon of light is used to photo decompose a hydrogen donor and the hydrogen being used to reduce CO2. Then in 1939, Robin Hill showed that isolated chloroplasts would make oxygen, but not fix CO2 showing the light and dark reactions occurred in different places. Although they are referred to as light and dark reactions, both of them take place only in the presence of light. This led later to the discovery of photosystems 1 and 2.

References

- Karr, James R. (1981). "Assessment of biotic integrity using fish communities". Fisheries. 6 (6): 21–27. doi:10.1577/1548-8446(1981)006<0021:AOBIUF>2.0.CO;2. ISSN 1548-8446.
- NCSU Water Quality Group. "Biomonitoring". WATERSHEDSS: A Decision Support System for Nonpoint Source Pollution Control. Raleigh, NC: North Carolina State University. Retrieved 2016-07-31.
- Tingey, David T. (1989). "Bio indicators in Air Pollution Research -- Applications and Constraints". Biologic Markers of Air-Pollution Stress and Damage in Forests. Washington, DC: National Academies Press: 73–80. ISBN 978-0-309-07833-7.
- Chessman, Bruce (2003). SIGNAL 2 – A Scoring System for Macro-invertebrate ('Water Bugs') in Australian Rivers (PDF). Monitoring River Heath Initiative Technical Report no. 31. Canberra: Commonwealth of Australia, Department of the Environment and Heritage. ISBN 0642548978.
- Reece J, Urry L, Cain M, Wasserman S, Minorsky P, Jackson R. Biology (International ed.). Upper Saddle River, NJ: Pearson Education. pp. 235, 244. ISBN 0-321-73975-2.
- "World Consumption of Primary Energy by Energy Type and Selected Country Groups, 1980–2004" (XLS). Energy Information Administration. July 31, 2006. Retrieved 2007-01-20.
- "Photosynthesis". McGraw-Hill Encyclopedia of Science & Technology. 13. New York: McGraw-Hill. 2007. ISBN 0-07-144143-3.
- Whitmarsh J, Govindjee (1999). "Chapter 2: The Basic Photosynthetic Process". In Singhal GS, Renger G, Sopory SK, Irrgang KD, Govindjee. Concepts in Photobiology: Photosynthesis and Photomorphogenesis. Boston: Kluwer Academic Publishers. p. 13. ISBN 978-0-7923-5519-9.

6

Green Chemistry: An Overview

Green chemistry is an important area of chemistry that focuses on designing products and also focuses on minimizing the use of hazardous substances. Some of the aspects explained are natural-gas processing, supercritical hydrolysis and condensation reaction. This chapter on green chemistry offers an overview on the subject matter.

Green Chemistry

Green chemistry, also called sustainable chemistry, is an area of chemistry and chemical engineering focused on the designing of products and processes that minimize the use and generation of hazardous substances. Whereas environmental chemistry focuses on the effects of polluting chemicals on nature, green chemistry focuses on technological approaches to preventing pollution and reducing consumption of nonrenewable resources.

Green chemistry overlaps with all subdisciplines of chemistry but with a particular focus on chemical synthesis, process chemistry, and chemical engineering, in industrial applications. To a lesser extent, the principles of green chemistry also affect laboratory practices. The overarching goals of green chemistry—namely, more resource-efficient and inherently safer design of molecules, materials, products, and processes—can be pursued in a wide range of contexts.

History

Green chemistry emerged from a variety of existing ideas and research efforts (such as atom economy and catalysis) in the period leading up to the 1990s, in the context of increasing attention to problems of chemical pollution and resource depletion. The development of green chemistry in Europe and the United States was linked to a shift in environmental problem-solving strategies: a movement from command and control regulation and mandated reduction of industrial emissions at the "end of the pipe," toward the active prevention of pollution through the innovative design of production technologies themselves. The set of concepts now recognized as green chemistry coalesced in the mid- to late-1990s, along with broader adoption of the term (which prevailed over competing terms such as "clean" and "sustainable" chemistry).

In the United States, the Environmental Protection Agency played a significant early role in fostering green chemistry through its pollution prevention programs, funding, and professional coordination. At the same time in the United Kingdom, researchers at the University of York contributed to the establishment of the Green Chemistry Network within the Royal Society of Chemistry.

Principles

In 1998, Paul Anastas (who then directed the Green Chemistry Program at the US EPA) and John C. Warner (then of Polaroid Corporation) published a set of principles to guide the practice of green chemistry. The twelve principles address a range of ways to reduce the environmental and health impacts of chemical production, and also indicate research priorities for the development of green chemistry technologies.

The principles cover such concepts as:

- the design of processes to maximize the amount of raw material that ends up in the product;
- the use of renewable material feedstocks and energy sources;
- the use of safe, environmentally benign substances, including solvents, whenever possible;
- the design of energy efficient processes;
- avoiding the production of waste, which is viewed as the ideal form of waste management.

The twelve principles of green chemistry are:

1. It is better to prevent waste than to treat or clean up waste after it is formed.
2. Synthetic methods should be designed to maximize the incorporation of all materials used in the process into the final product.
3. Wherever practicable, synthetic methodologies should be designed to use and generate substances that possess little or no toxicity to human health and the environment.
4. Chemical products should be designed to preserve efficacy of function while reducing toxicity.
5. The use of auxiliary substances (e.g. solvents, separation agents, etc.) should be made unnecessary wherever possible and innocuous when used.
6. Energy requirements should be recognized for their environmental and economic impacts and should be minimized. Synthetic methods should be conducted at ambient temperature and pressure.

7. A raw material or feedstock should be renewable rather than depleting wherever technically and economically practicable.
8. Reduce derivatives – Unnecessary derivatization (blocking group, protection/ deprotection, temporary modification) should be avoided whenever possible.
9. Catalytic reagents (as selective as possible) are superior to stoichiometric reagents.
10. Chemical products should be designed so that at the end of their function they do not persist in the environment and break down into innocuous degradation products.
11. Analytical methodologies need to be further developed to allow for real-time, in-process monitoring and control prior to the formation of hazardous substances.
12. Substances and the form of a substance used in a chemical process should be chosen to minimize potential for chemical accidents, including releases, explosions, and fires.

Trends

Attempts are being made not only to quantify the *greenness* of a chemical process but also to factor in other variables such as chemical yield, the price of reaction components, safety in handling chemicals, hardware demands, energy profile and ease of product workup and purification. In one quantitative study, the reduction of nitrobenzene to aniline receives 64 points out of 100 marking it as an acceptable synthesis overall whereas a synthesis of an amide using HMDS is only described as adequate with a combined 32 points.

Green chemistry is increasingly seen as a powerful tool that researchers must use to evaluate the environmental impact of nanotechnology. As nanomaterials are developed, the environmental and human health impacts of both the products themselves and the processes to make them must be considered to ensure their long-term economic viability.

Examples

Green Solvents

Solvents are consumed in large quantities in many chemical syntheses as well as for cleaning and degreasing. Traditional solvents are often toxic or are chlorinated. Green solvents, on the other hand, are generally derived from renewable resources and biodegrade to innocuous, often naturally occurring product.

Synthetic Techniques

Novel or enhanced synthetic techniques can often provide improved environmental performance or enable better adherence to the principles of green chemistry. For example, the 2005 Nobel Prize for Chemistry was awarded, to Yves Chauvin, Robert H. Grubbs and Richard R. Schrock, for the development of the metathesis method in organic synthesis, with explicit reference to its contribution to green chemistry and "smarter production." A 2005 review identified three key developments in green chemistry in the field of organic synthesis: use of supercritical carbon dioxide as green solvent, aqueous hydrogen peroxide for clean oxidations and the use of hydrogen in asymmetric synthesis. Some further examples of applied green chemistry are supercritical water oxidation, on water reactions, and dry media reactions.

Bioengineering is also seen as a promising technique for achieving green chemistry goals. A number of important process chemicals can be synthesized in engineered organisms, such as shikimate, a Tamiflu precursor which is fermented by Roche in bacteria. Click chemistry is often cited as a style of chemical synthesis that is consistent with the goals of green chemistry. The concept of 'green pharmacy' has recently been articulated based on similar principles.

Carbon Dioxide as Blowing Agent

In 1996, Dow Chemical won the 1996 Greener Reaction Conditions award for their 100% carbon dioxide blowing agent for polystyrene foam production. Polystyrene foam is a common material used in packing and food transportation. Seven hundred million pounds are produced each year in the United States alone. Traditionally, CFC and other ozone-depleting chemicals were used in the production process of the foam sheets, presenting a serious environmental hazard. Flammable, explosive, and, in some cases toxic hydrocarbons have also been used as CFC replacements, but they present their own problems. Dow Chemical discovered that supercritical carbon dioxide works equally as well as a blowing agent, without the need for hazardous substances, allowing the polystyrene to be more easily recycled. The CO_2 used in the process is reused from other industries, so the net carbon released from the process is zero.

Hydrazine

Addressing principle #2 is the Peroxide Process for producing hydrazine without cogenerating salt. Hydrazine is traditionally produced by the Olin Raschig process from sodium hypochlorite (the active ingredient in many bleaches) and ammonia. The net reaction produces one equivalent of sodium chloride for every equivalent of the targeted product hydrazine:

$$NaOCl + 2\,NH_3 \rightarrow H_2N\text{-}NH_2 + NaCl + H_2O$$

In the greener Peroxide process hydrogen peroxide is employed as the oxidant, the side product being water. The net conversion follows:

$$2\,NH_3 + H_2O_2 \rightarrow H_2N\text{-}NH_2 + 2\,H_2O$$

Addressing principle #4, this process does not require auxiliary extracting solvents. Methyl ethyl ketone is used as a carrier for the hydrazine, the intermediate ketazide phase separates from the reaction mixture, facilitating workup without the need of an extracting solvent.

1,3-Propanediol

Addressing principle #7 is a green route to 1,3-propanediol, which is traditionally generated from petrochemical precursors. It can be produced from renewable precursors via the bioseparation of 1,3-propanediol using a genetically modified strain of *E. coli*. This diol is used to make new polyesters for the manufacture of carpets.

Lactide

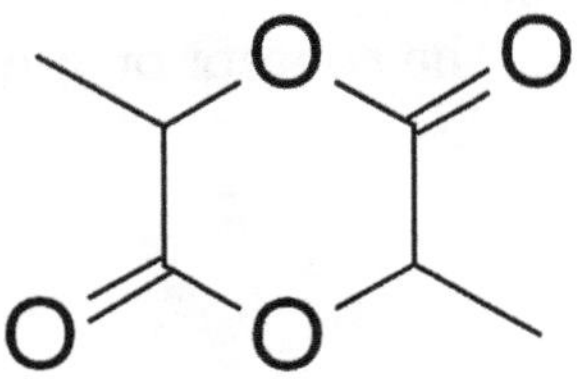

Lactide

In 2002, Cargill Dow (now NatureWorks) won the Greener Reaction Conditions Award for their improved method for polymerization of polylactic acid . Unfortunately, lactide-base polymers do not perform well and the project was discontinued by Dow soon after the award. Lactic acid is produced by fermenting corn and converted to lactide, the cyclic dimer ester of lactic acid using an efficient, tin-catalyzed cyclization. The L,L-lactide enantiomer is isolated by distillation and polymerized in the melt to make a crystallizable polymer, which has some applications including textiles and apparel, cutlery, and food packaging. Wal-Mart has announced that it is using/will use PLA for its produce packaging. The NatureWorks PLA process substitutes renewable materials for petroleum feedstocks, doesn't require the use of hazardous organic solvents typical in other PLA processes, and results in a high-quality polymer that is recyclable and compostable.

Carpet Tile Backings

In 2003 Shaw Industries selected a combination of polyolefin resins as the base polymer of choice for EcoWorx due to the low toxicity of its feedstocks, superior adhesion properties, dimensional stability, and its ability to be recycled. The EcoWorx compound also had to be designed to be compatible with nylon carpet fiber. Although EcoWorx may be recovered from any fiber type, nylon-6 provides a significant advantage. Polyolefins are compatible with known nylon-6 depolymerization methods. PVC interferes with those processes. Nylon-6 chemistry is well-known and not addressed in first-generation production. From its inception, EcoWorx met all of the design criteria necessary to satisfy the needs of the marketplace from a performance, health, and environmental standpoint. Research indicated that separation of the fiber and backing through elutriation, grinding, and air separation proved to be the best way to recover the face and backing components, but an infrastructure for returning postconsumer EcoWorx to the elutriation process was necessary. Research also indicated that the postconsumer carpet tile had a positive economic value at the end of its useful life. EcoWorx is recognized by MBDC as a certified cradle-to-cradle design.

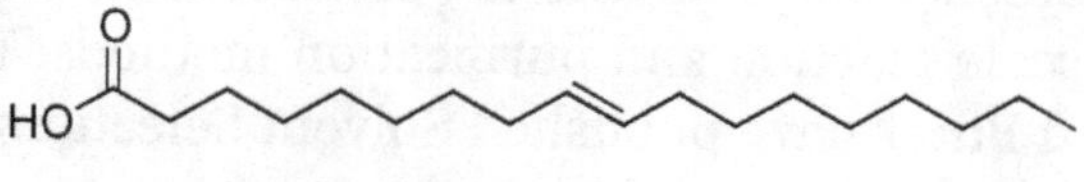

trans-Oleic acid

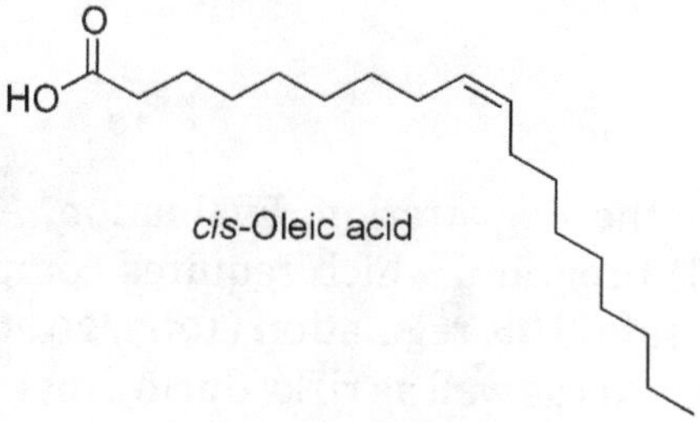

cis-Oleic acid

Trans and *cis* fatty acids

Transesterification of Fats

In 2005, Archer Daniels Midland (ADM) and Novozymes won the Greener Synthetic Pathways Award for their enzyme interesterification process. In response to the U.S. Food and Drug Administration (FDA) mandated labeling of *trans*-fats on nutritional information by January 1, 2006, Novozymes and ADM worked together to develop a clean, enzymatic process for the interesterification of oils and fats by interchanging saturated and unsaturated fatty acids. The result is commercially viable products without *trans*-fats. In addition to the human health benefits of eliminating *trans*-fats, the process has reduced the use of toxic chemicals and water, prevents vast amounts of byproducts, and reduces the amount of fats and oils wasted.

Bio-succinic Acid

In 2011, the Outstanding Green Chemistry Accomplishments by a Small Business Award went to BioAmber Inc. for integrated production and downstream applications of bio-based succinic acid. Succinic acid is a platform chemical that is an important starting material in the formulations of everyday products. Traditionally, succinic acid is produced from petroleum-based feedstocks. BioAmber has developed process and technology that produces succinic acid from the fermentation of renewable feedstocks at a lower cost and lower energy expenditure than the petroleum equivalent while sequestering CO_2 rather than emitting it.

Laboratory Chemicals

Several laboratory chemicals are controversial from the perspective of Green chemistry. The Massachusetts Institute of Technology has created the to help identify alternatives. Ethidium bromide, xylene, mercury, and formaldehyde have been identified as "worst offenders" which have alternatives. Solvents in particular make a large contribution to the environmental impact of chemical manufacturing and there is a growing focus on introducing Greener solvents into the earliest stage of development of these processes: laboratory-scale reaction and purification methods. In the Pharmaceutical Industry, both GSK and Pfizer have published Solvent Selection Guides for their Drug Discovery chemists.

Legislation

The EU

In 2007, The EU put into place the Registration, Evaluation, Authorisation, and Restriction of Chemicals (REACH) program, which requires companies to provide data showing that their products are safe. This regulation (1907/2006) ensures not only the assessment of the chemicals' hazards as well as risks during their uses but also includes measures for banning or restricting/authorising uses of specific substances. ECHA, the EU Chemicals Agency in Helsinki, is implementing the regulation whereas the enforcement lies with the EU member states.

United States

The U.S. law that governs the majority of industrial chemicals (excluding pesticides, foods, and pharmaceuticals) is the Toxic Substances Control Act (TSCA) of 1976. Examining the role of regulatory programs in shaping the development of green chemistry in the United States, analysts have revealed structural flaws and long-standing weaknesses in TSCA; for example, a 2006 report to the California Legislature concludes that TSCA has produced a domestic chemicals market that discounts the hazardous properties of chemicals relative to their function, price, and performance. Scholars have argued that such market conditions represent a key barrier to the scientific, technical, and commercial success of green chemistry in the U.S., and fundamental policy changes are needed to correct these weaknesses.

Passed in 1990, the Pollution Prevention Act helped foster new approaches for dealing with pollution by preventing environmental problems before they happen.

In 2008, the State of California approved two laws aiming to encourage green chemistry, launching the California Green Chemistry Initiative. One of these statutes required California's Department of Toxic Substances Control (DTSC) to develop new regulations to prioritize "chemicals of concern" and promote the substitution of hazardous chemicals with safer alternatives. The resulting regulations took effect in 2013, initiating DTSC's *Safer Consumer Products Program*.

Green Chemistry Education

Many institutions offer courses and degrees on Green Chemistry. Examples from across the globe are Denmark's Technical University, and several in the US, e.g. at the Universities of Massachusetts-Boston, Michigan, and Oregon. A masters level course in Green Technology, has been introduced by the Institute of Chemical Technology, India. In the UK at the University of York University of Leicester, Department of Chemistry and MRes in Green Chemistry at Imperial College London. In Spain different universities like the Universidad de Jaume I or the Universidad de Navarra, offer Green Chemistry master courses. There are also websites focusing on green chemistry, such as the Michigan Green Chemistry Clearinghouse at www.migreenchemistry.org. Apart from its Green Chemistry Master courses the Zurich University of Applied Sciences ZHAW presents an exposition and web page "Making chemistry green" for a broader public, illustrating the 12 principles.

Contested Definition

There are ambiguities in the definition of green chemistry, and in how it is understood among broader science, policy, and business communities. Even within chemistry, researchers have used the term "green chemistry" to describe a range of work independently of the framework put forward by Anastas and Warner (i.e., the 12 principles). While not all uses of the term are legitimate, many are, and the authoritative status of any single definition is uncertain. More broadly, the idea of green chemistry can easily be linked (or confused) with related concepts like green engineering, environmental design, or sustainability in general. The complexity and multifaceted nature of green chemistry makes it difficult to devise clear and simple metrics. As a result, "what is green" is often open to debate.

Green Chemistry Awards

Several scientific societies have created awards to encourage research in green chemistry.

- Australia's Green Chemistry Challenge Awards overseen by The Royal Australian Chemical Institute (RACI).
- The Canadian Green Chemistry Medal.
- In Italy, Green Chemistry activities center around an inter-university consortium known as INCA.
- In Japan, The Green & Sustainable Chemistry Network oversees the GSC awards program.

- In the United Kingdom, the Green Chemical Technology Awards are given by Crystal Faraday.
- In the US, the Presidential Green Chemistry Challenge Awards recognize individuals and businesses.

Natural-gas Processing

A natural-gas processing plant

Natural-gas processing is a complex industrial process designed to clean raw natural gas by separating impurities and various non-methane hydrocarbons and fluids to produce what is known as *pipeline quality* dry natural gas.

Natural-gas processing begins at the well head. The composition of the raw natural gas extracted from producing wells depends on the type, depth, and location of the underground deposit and the geology of the area. Oil and natural gas are often found together in the same reservoir. The natural gas produced from oil wells is generally classified as *associated-dissolved*, meaning that the natural gas is associated with or dissolved in crude oil. Natural gas production absent any association with crude oil is classified as "non-associated." In 2009, 89 percent of U.S. wellhead production of natural gas was non-associated.

Natural-gas processing plants purify raw natural gas by removing common contaminants such as water, carbon dioxide (CO_2) and hydrogen sulfide (H_2S). Some of the substances which contaminate natural gas have economic value and are further processed or sold. A fully operational plant delivers pipeline-quality dry natural gas that can be used as fuel by residential, commercial and industrial consumers.

Types of Raw-natural-gas Wells

Raw natural gas comes primarily from any one of three types of wells: crude oil wells, gas wells, and condensate wells.

Natural gas that comes from crude oil wells is typically called *associated gas*. This gas can have existed as a gas cap above the crude oil in the underground formation, or could have been dissolved in the crude oil.

Natural gas from gas wells and from condensate wells, in which there is little or no crude oil, is called *non-associated gas*. Gas wells typically produce only raw natural gas, while condensate wells produce raw natural gas along with other low molecular weight hydrocarbons. Those that are liquid at ambient conditions (i.e., pentane and heavier) are called *natural gas condensate* (sometimes also called *natural gasoline* or simply *condensate*).

Natural gas is called *sweet gas* when relatively free of hydrogen sulfide; gas that does contain hydrogen sulfide is called *sour gas*. Natural gas, or any other gas mixture, containing significant quantities of hydrogen sulfide, carbon dioxide or similar acidic gases, is called *acid gas*

Raw natural gas can also come from methane deposits in the pores of coal seams, and especially in a more concentrated state of adsorption onto the surface of the coal itself. Such gas is referred to as *coalbed gas* or *coalbed methane* (*coal seam gas* in Australia). Coalbed gas has become an important source of energy in recent decades.

Contaminants in Raw Natural Gas

Raw natural gas typically consists primarily of methane (CH_4), the shortest and lightest hydrocarbon molecule. It also contains varying amounts of:

- Heavier gaseous hydrocarbons: ethane (C_2H_6), propane (C_3H_8), normal butane (n-C_4H_{10}), isobutane (i-C_4H_{10}), pentanes and even higher molecular weight hydrocarbons. When processed and purified into finished by-products, all of these are collectively referred to as Natural Gas Liquids or NGL.
- Acid gases: carbon dioxide (CO_2), hydrogen sulfide (H_2S) and mercaptans such as methanethiol (CH_3SH) and ethanethiol (C_2H_5SH).
- Other gases: nitrogen (N_2) and helium (He).
- Water: water vapor and liquid water. Also dissolved salts and dissolved gases (acids).
- Liquid hydrocarbons: perhaps some natural-gas condensate (also referred to as *casinghead gasoline* or *natural gasoline*) and/or crude oil.

- Mercury: very small amounts of mercury primarily in elemental form, but chlorides and other species are possibly present.
- Naturally occurring radioactive material (NORM): natural gas may contain radon, and the produced water may contain dissolved traces of radium, which can accumulate within piping and processing equipment. This can render piping and equipment radioactive over time.

The raw natural gas must be purified to meet the quality standards specified by the major pipeline transmission and distribution companies. Those quality standards vary from pipeline to pipeline and are usually a function of a pipeline system's design and the markets that it serves. In general, the standards specify that the natural gas:

- Be within a specific range of heating value (caloric value). For example, in the United States, it should be about 1035 ± 5% BTU per cubic foot of gas at 1 atmosphere and 60°F (41 MJ ± 5% per cubic metre of gas at 1 atmosphere and 15.6°C).
- Be delivered at or above a specified hydrocarbon dew point temperature (below which some of the hydrocarbons in the gas might condense at pipeline pressure forming liquid slugs that could damage the pipeline).
- Dew-point adjustment serves the reduction of the concentration of water and heavy hydrocarbons in natural gas to such an extent that no condensation occurs during the ensuing transport in the pipelines
- Be free of particulate solids and liquid water to prevent erosion, corrosion or other damage to the pipeline.
- Be dehydrated of water vapor sufficiently to prevent the formation of methane hydrates within the gas processing plant or subsequently within the sales gas transmission pipeline. A typical water content specification in the U.S. is that gas must contain no more than seven pounds of water per million standard cubic feet (MMSCF) of gas.
- Contain no more than trace amounts of components such as hydrogen sulfide, carbon dioxide, mercaptans, and nitrogen. The most common specification for hydrogen sulfide content is 0.25 grain H_2S per 100 cubic feet of gas, or approximately 4 ppm. Specifications for CO_2 typically limit the content to no more than two or three percent.
- Maintain mercury at less than detectable limits (approximately 0.001 ppb by volume) primarily to avoid damaging equipment in the gas processing plant or the pipeline transmission system from mercury amalgamation and embrittlement of aluminum and other metals.

Description of a Natural-gas Processing Plant

There are a great many ways in which to configure the various unit processes used in the processing of raw natural gas. The block flow diagram below is a generalized, typical configuration for the processing of raw natural gas from non-associated gas wells. It shows how raw natural gas is processed into sales gas pipelined to the end user markets. It also shows how processing of the raw natural gas yields these byproducts:

- Natural-gas condensate
- Sulfur
- Ethane
- Natural-gas liquids (NGL): propane, butanes and C_5+ (which is the commonly used term for pentanes plus higher molecular weight hydrocarbons)

Raw natural gas is commonly collected from a group of adjacent wells and is first processed at that collection point for removal of free liquid water and natural gas condensate. The condensate is usually then transported to an oil refinery and the water is disposed of as wastewater.

The raw gas is then pipelined to a gas processing plant where the initial purification is usually the removal of acid gases (hydrogen sulfide and carbon dioxide). There are many processes that are available for that purpose, but amine treating is the process that was historically used. However, due to a range of performance and enviro-nmental constraints of the amine process, a newer technology based on the use of polymeric membranes to separate the carbon dioxide and hydrogen sulfide from the natural gas stream has gained increasing acceptance. Membranes are attractive since no reagents are consumed.

The acid gases, if present, are removed by membrane or amine treating can then be routed into a sulfur recovery unit which converts the hydrogen sulfide in the acid gas into either elemental sulfur or sulfuric acid. Of the processes available for these conversions, the Claus process is by far the most well known for recovering elemental sulfur, whereas the conventional Contact process and the WSA (Wet sulfuric acid process) are the most used technologies for recovering sulfuric acid.

The residual gas from the Claus process is commonly called *tail gas* and that gas is then processed in a tail gas treating unit (TGTU) to recover and recycle residual sulfur-containing compounds back into the Claus unit. Again, there are a number of processes available for treating the Claus unit tail gas and for that purpose a WSA process is also very suitable since it can work autothermally on tail gases.

The next step in the gas processing plant is to remove water vapor from the gas using either the regenerable absorption in liquid triethylene glycol (TEG), commonly referred to as glycol dehydration, deliquescent chloride desiccants, and or a Pressure Swing Adsorption (PSA) unit which is regenerable adsorption using a solid adsorbent. Other newer processes like membranes may also be considered.

Mercury is then removed by using adsorption processes such as activated carbon or regenerable molecular sieves.

Although not common, nitrogen is sometimes removed and rejected using one of the three processes indicated on the flow diagram:

- Cryogenic process (Nitrogen Rejection Unit), using low temperature distillation. This process can be modified to also recover helium, if desired.
- Absorption process, using lean oil or a special solvent as the absorbent.
- Adsorption process, using activated carbon or molecular sieves as the adsorbent. This process may have limited applicability because it is said to incur the loss of butanes and heavier hydrocarbons.

The next step is to recover the natural gas liquids (NGL) for which most large, modern gas processing plants use another cryogenic low temperature distillation process involving expansion of the gas through a turbo-expander followed by distillation in a demethanizing fractionating column. Some gas processing plants use lean oil absorption process rather than the cryogenic turbo-expander process.

The residue gas from the NGL recovery section is the final, purified sales gas which is pipelined to the end-user markets.

The recovered NGL stream is sometimes processed through a fractionation train consisting of three distillation towers in series: a deethanizer, a depropanizer and a debutanizer. The overhead product from the deethanizer is ethane and the bottoms are fed to the depropanizer. The overhead product from the depropanizer is propane and the bottoms are fed to the debutanizer. The overhead product from the debutanizer is a mixture of normal and iso-butane, and the bottoms product is a C_5+ mixture. The recovered streams of propane, butanes and C_5+ may be "sweetened" in a Merox process unit to convert undesirable mercaptans into disulfides and, along with the recovered ethane, are the final NGL by-products from the gas processing plant. Currently, most cryogenic plants do not include fractionation for economic reasons, and the NGL stream is instead transported as a mixed product to standalone fractionation complexes located near refineries or chemical plants that use the components for feedstock. In case laying pipeline is not possible for geographical reason,or the distance between source and consumer exceed 3000 km, natural gas is then transported by ship as LNG (liquefied natural gas) and again converted into its gaseous state in the vicinity of the consumer.

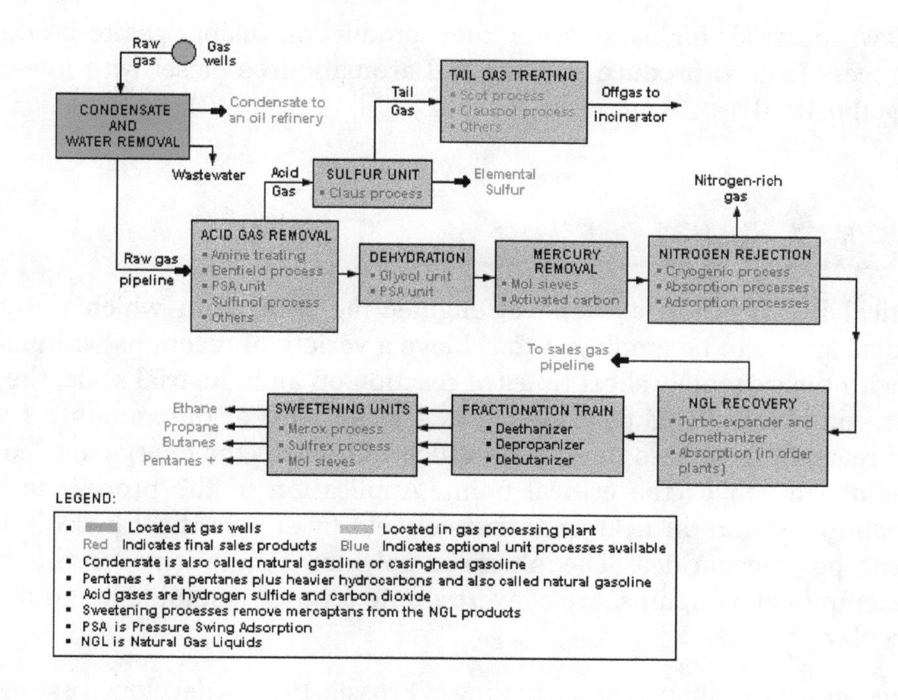

Helium Recovery

If the gas contains significant helium content, the helium may be recovered by fractional distillation. Natural gas may contain as much as 7% helium, and is the commercial source of the noble gas. For instance, the Hugoton Gas Field in Kansas and Oklahoma in the United States contains concentrations of helium from 0.3% to 1.9%, which is separated out as a valuable byproduct.

Consumption

Natural gas consumption patterns, across nations, vary based on access. Countries with large reserves tend to handle the raw-material natural gas more generously, while countries with scarce or lacking resources tend to be more economical. Despite the considerable findings, the predicted availability of the natural-gas reserves has hardly changed.

Applications of Natural Gas

- Fuel for industrial heating and desiccation process
- Fuel for the operation of public and industrial power stations
- Household fuel for cooking, heating and providing hot water
- Fuel for environmentally friendly compressed or liquid natural gas vehicles
- Raw material for chemical synthesis

- Raw material for large-scale fuel production using gas-to-liquid (GTL) process (e.g. to produce sulphur-and aromatic-free diesel with low-emission combustion)

Supercritical Hydrolysis

Supercritical hydrolysis is a chemical engineering process in which water in the supercritical state can be employed to achieve a variety of reactions within seconds. To cope with the extremely short times of reaction on an industrial scale, the process should be continuous. This continuity enables the ratio of the amount of water to the other reactant to be less than unity which minimizes the energy needed to heat the water above 374 C, the critical point. Application of the process to biomass provides simple sugars in near quantitative yield by supercritical hydrolysis of the constituent polysaccharides. The phenolic polymer components of the biomass, usually exemplified by lignins, are converted into a water-insoluble liquid mixture of low molecular phenols.

A private company, Renmatix, based in King of Prussia, PA, has developed a supercritical hydrolysis technology to convert a range of non-food biomass feedstocks into cellulosic sugars for application in biochemicals and biofuels. It has a demonstration facility in Georgia, currently capable of processing three dry tons of hardwood biomass into cellulosic sugar daily. In Australia, a government-sponsored entity called Licella, is similarly transforming sawdust. Both processes require high ratios of water to the amount of feedstock. This energy profligacy can be avoided by the use of a plastic-type extruder through which the solid, but wet, biomass is conveyed to a small inductively heated reaction zone.

Supercritical hydrolysis can be considered a broadly applicable green chemistry process that utilizes water simultaneously as a heat transfer agent, a solvent, a reactant, a source of hydrogen and as a char-reduction component.

California Green Chemistry Initiative

The California Green Chemistry Initiative (CGCI) is a six-part initiative to reduce public and environmental exposure to toxins through improved knowledge and regulation of chemicals; two parts became statute in 2008. The other four parts were not passed, but are still on the agenda of the California Department of Toxic Substances Control green ribbon science panel discussions. The two parts of the California Green Chemistry Initiative that were passed are known as AB 1879 (Chapter 559, Statutes of 2008): Hazardous Materials and Toxic Substances Evaluation and Regulation and SB 509

(Chapter 560, Statutes of 2008): Toxic Information Clearinghouse. Implementation of CGCI has been delayed indefinitely beyond the January 1, 2011.

Purpose

Green chemistry is the design of chemical products and processes that reduce or eliminate the use and generation of hazardous substances. Green chemistry is based upon twelve principles, identified in "Green Chemistry: Theory and Practice" and adopted by the US Environmental Protection Agency (EPA). It is an innovative technology which encourages the design of safer chemicals and products and minimizes the impact of wastes through increased energy efficiency, the design of chemical products that degrade after use and the use of renewable resources (instead of non-renewable fossil fuel such as petroleum, gas and coal). The Office of Pollution Prevention and Toxics (OPPT), created under the United States Pollution Prevention Act of 1990, promotes the use of chemistry for pollution prevention through voluntary, non-regulatory ' partnerships with academia, industry, other government agencies, and non-governmental organizations. The United States Environmental Protection Agency (EPA) promotes green chemistry as overseen by the OPPT. The California Green Chemistry Initiative moves beyond voluntary partnerships and voluntary information disclosure to require industry reporting and public disclosure.

Overview

The United States Environmental Protection Agency's most important law to regulate the production, use and disposal of chemicals is the Toxic Substances Control Act of 1976 (TSCA). Over the years, TSCA has fallen behind the industry it is supposed to regulate and is an inadequate tool for providing the protection against today's chemical risks. Green chemistry represents a major paradigm shift in industrial manufacturing as it is a proactive "cradle-to-cradle" approach that focuses environmental protection at the design stage of production processes.

In 2008, California governor Arnold Schwarzenegger signed two joined bills, AB 1879 and SB 507, which created California's Green Chemistry Initiative (CGCI). AB 1879 increases regulatory authority over chemicals in consumer products. The law established an advisory panel of scientists, known as the green ribbon science panel, to guide research in chemical policy, create regulations for assessing alternatives, and set up an internet database of research on toxins. SB 509 was designed to ensure that information regarding the hazard traits, toxicological and environmental endpoints, and other vital data is available to the public, to businesses, and to regulators in a Toxics Information Clearinghouse. This legislation marks the biggest leap forward in California chemicals policy in nearly two decades and is intended to improve the health and safety of all Californians by providing the Department of Toxic Substances Control (DTSC) with the authority to control toxic substances in consumer products.

The bills were scheduled to go into regulatory affect January 1, 2011 with the adoption of the Green Chemistry Initiative. California has postponed the initiative, indefinitely, due to concerns raised by stakeholders and more specifically, controversial last minute changes in the final draft. The final or third draft contains substantial revisions, including scaled back manufacturer and retailer compliance requirements that were not well received by the environmental community. Assemblyman Mike Feur and several authors of AB 1879, assert that last minute changes by the California DTSC have drastically weakened the Green Chemistry Initiative and limited its scope. They are most concerned with the change to require the state to prove that a chemical is harmful before being regulated, mirroring what is currently required at the Federal level by TSCA. The original draft advocated a precautionary principle, or "cradle-to-cradle" approach. Environmentalists fear that CGCI will not remove chemicals off the shelves, but instead will create "paralysis by analysis" as companies litigate against the DTSC over unfavorable decisions.

Physical and Social Causes

Traditional Methods of Dealing with Wastes

Society historically managed its industrial and municipal wastes by disposal or incineration. Chemical regulation occurs only after a product is identified as hazardous. This problem-specific approach has led to the release of thousands of potentially harmful chemicals in our environment. Chemical regulation is a continuous game of catch up, in which banned chemicals are replaced with new chemicals that may be just as or more toxic. Many environmental laws are still based on the industrial production model of cradle-to-grave. The term "cradle-to-grave" is used to describe and assess the life-cycle of products, from raw material extraction through materials processing, manufacture, distribution, use and disposal. This traditional approach to chemicals management has serious environmental drawbacks because it does not consider what happens to a product after it is disposed of. The Resource Conservation and Recovery Act (RCRA) of 1976, exemplifies a cradle-to-grave management approach of hazardous waste. RCRA has been largely ineffective because its emphasis is on dealing with waste after it has been created; meanwhile emphasis on waste reduction is minimal. Waste does not disappear, it is simply transported elsewhere. Costly and burdensome hazardous waste disposal in the US has encouraged the exportation of hazardous waste to poor counties and developing nations willing to accept the waste for a fee.

The Green Chemistry initiative instead employs a cradle-to-cradle approach, representing a major paradigm shift in environmental policy and provides a proactive solution to toxic waste. The Earth's capacity to accept toxic waste is practically nonexistent. The disposal of hazardous wastes is not the root problem but rather, the root symptom. The critical issue is the creation of toxic wastes. Requiring manufacturers to consider chemical exposure during manufacturing, throughout product use and after disposal, encourages the production of safer products.

Consumption and Wastes

By the time we find a product on a market shelf, 90% of the resources used to create that product was regarded as waste. This accounts for about 136 pounds of resources a week consumed by the average American and 2,000 pounds of waste support that consumption. As the population grows and the economy expands more and more products will be created, consumed, and disposed. Many negative externalities are related to the environmental consequences of production and use, including air pollution, anthropogenic climate change and water pollution. Under the current cycle of production, toxic chemical byproducts will continue to be produced and unleashed on our environment. It is important to carefully consider how toxic wastes are created in order to forgo the possibility of a world that is unsuitable for human life.

Transparency Issues

One of the biggest failures in market transactions is the imbalance of information that is provided to consumer via producer. "Information asymmetry" is an economic concept that is used to explain this failure: it deals with the study of decisions in transactions where one party has more or better information than the other. Due to a lack of information transparency, the public may lack vital information about the health and safety of products found on supermarket shelves. This lack of information may have led to a reversed purchasing decision. Yet without such labeling, consumers must make assumptions based on things like price or expertise. For example, one apple juice brand may be assumed healthier because it cost more and because the brand is advertised as "healthy" and "recommended by mothers". Further, it may be assumed that the product is safe for consumption if it is sitting on a grocery store shelf and probably would not be approved by the government if it contained harmful chemicals. Assumptions such as these could inform a typical purchasing decision, despite their inaccuracy. Perhaps given more information, the same brand of apple juice would be less desirable if information on unhealthy preservatives, additives or pesticide residues was easily obtained. To make market transactions more efficient, the government could force more accurate labeling about products, laws could require companies to be more transparent, and the government could require that advertising be less persuasive and more informative. The Green Chemistry Initiative of California would address transparency issues by creating a public chemical inventory and requiring more stringent regulation of chemicals that may be toxic. The CGCI Draft Report suggests a green labeling system to identify consumer products with ingredients harmful to human health and the environment.

Stakeholder Involvement

The United States is the world leader in chemicals manufacturing. As a multibillion-dollar industry, the chemical industry has a leading role in the US economy and because of this, a high level of influence in federal decision-making. Central to the modern world economy, it converts raw materials (oil, natural gas, air, water, metals, and minerals) into more than 70,000 different products. The chemical industry—producers of chemicals, household cleansers, plastics, rubber, paints and explosives, keeps a watchful eye on issues including environmental and health policy, taxes and trade. The industry is often the target of environmental groups, which charge that chemicals and chemical waste are polluting the air and water supply. And like most industries with pollution problems, chemical manufacturers oppose meddlesome government regulations that make it more difficult and expensive for them to do business. So do most Republicans, which is why this industry gives nearly three-fourths of its campaign contributions to the GOP. In addition to campaign contributions to elected officials and candidates, companies, labor unions, and other organizations spend billions of dollars each year to lobby Congress and federal agencies. Some special interests retain lobbying firms, many of them located along Washington's legendary K Street; others have lobbyists working in-house.

According to website *Opensecrets,* the total number of clients lobbying for the chemical industry in 2010 was 143, which is the highest number in history. The first group on this list, American Chemistry Council spent $8,130,000 lobbying last year and Crop America, which comes second, spent $2,291,859 lobbying last year, FMC Corporation spent $1,230,000 and Koch Industries spent $8,070,000. The Chemical Industry wants limited testing of chemicals, more lengthy and costly studies of chemicals already proven to be dangerous, and an assumption that we are only exposed to one chemical at a time, and from one source at a time.

According to *Safer Chemicals, Healthy Families,* a broad coalition of groups, including major environmental organizations like the Natural Resources Defense Council and the Environmental Defense Fund, health organizations like the Learning Disabilities Association, Breast Cancer Fund, and the Autism Society of America, health professionals and providers like the American Nurses Association, Planned Parenthood Federation of America, and the Mt. Sinai Children's Environmental Health Center, and concerned parents groups like MomsRising: there is growing national momentum and pressure to change the Toxic Substances Control Act (TSCA), our federal system for overseeing chemical safety, which has not been updated in thirty-five years. Polling data indicates overwhelming support for chemical regulation nationwide. According to polling data conducted by the Mellman Group, 84% say that "tightening controls" on chemical regulation is important, with 50% of those calling it "very important." Public Health Advocates want public disclosure of safety information for all chemicals in use, prompt action to phase out or reduce the most dangerous chemicals, deciding safety based on real world exposure to all sources of toxic chemicals.

History

In 2008, California Governor Arnold Schwarzenegger signed two state bills authorizing the state to identify toxic chemicals in industry and consumer products and analyze alternatives. AB 1879, written by Assemblyman Mike Feur, a Los Angeles Democrat, requires the state Department of Toxic Substances Control to assess chemicals and prioritize the most toxic for possible restrictions or bans. The environmental policy council, made up of heads of all state environmental protection agency boards and departments will oversee the program. SB 509, by Senator Joe Simitian, a Palo Alto Democrat, creates an online toxics information clearinghouse with information about the hazards of thousands of chemicals used in California. These bills are intended to put an end to chemical-by-chemical bans and remove harmful products at the design stage. The regulations are expected to motivate manufacturers of consumer products containing chemicals of concern to seek safer alternatives.

Supporters of the bill include the California Association of Professional Scientists, the Chemical Industry Council of California, DuPont, BIOCOM, Grocery Manufacturers Association, the Breast Cancer Fund, Catholic Healthcare West, in addition to a broad array of environmental groups such as the Coalition for Clean Air, the Environmental Defense Fund, the Natural Resources Defense Council. The American Electronics Association (AEA) and Ford spoke in opposition to the bill, each requesting an exemption from its provisions. Also opposing were environmental justice advocates who indicated the bill did not go far enough. Meanwhile, large trade associations such as Consumer Specialty Products Association, Western States Petroleum Association, American Chemistry Council, CA Manufacturers and Technology Association, and CA Chamber of Commerce officially withdrew opposition to the measures.

Due to outdated and inefficient or otherwise voluntary chemical regulation at the Federal level, the State of California has decided to take regulation into its own hands and develop stricter, environmentally-informed methodologies for dealing with the production of toxic wastes. California's economy is the largest of any state in the US, and is the eighth largest economy in the world. This position gives California an advantage when it comes to environmental standards: the impact of chemical regulation statewide can have a broader impact nationwide if manufacturers desire to stay competitive in California's market. The Green Chemistry Initiative forces statewide industries to comply with greener standards of production, which may spark innovation on a wider basis.

The Green Chemistry initiative aims to regulate the creation and use of materials hazardous to human health and the environment by encouraging innovative design and manufacturing, and ultimately safer consumer product alternatives. To develop the regulatory framework, DTSC held a number of stakeholder and public workshops and invited direct public participation in the drafting of regulations on a wiki website. DTSC reportedly received over 57,000 comments and over 800 regulatory suggestions. Regulatory suggestions included industry assessments of risk and safety, alternative chemicals and life-cycle assessments and mandatory industry reporting, full public disclosure of substances contained in products, a green labelling program that would inform consumers of the potential health and environmental impacts of the chemicals contained in products and a mandated surcharge on chemicals and products to support a fund to address environmental problems. In December 2008, DTSC announced six policy recommendations for the Green Chemistry Initiative. In brief, those recommendations are:

1. expand pollution prevention
2. develop green chemistry workforce education and training, research and development, technology transfer
3. online product ingredient network
4. online toxics clearing house
5. accelerate the quest for safer products
6. move toward cradle to cradle economy

Two of the six recommendations from this report were adopted: AB 1879 requires the DTSC to implement regulations to identify and prioritize chemicals of concern, evaluate alternatives, and specify regulatory responses where chemicals are found in products. SB 509 requires an online, public toxics information clearinghouse that includes science-based information on the toxicity and hazard traits of chemicals used in daily life. Essentially the recommended policy methods include authority tools that would regulate the approval on new chemicals in a more cautious manner as well as mandate the decimation of information, as provided by manufacturers to the public; innovation would be encouraged under this paradigm to replace harmful chemicals with greener alternatives and the California government would fund programs to help industries produce greener chemicals. Secondly, capacity or learning tools would be provided to the public in the form of the online database, giving the tools so that they have better ability to make market decisions that reflect their interests.

Criticism

Environmentalists say the amended regulations won't remove toxic products from the shelves and will create "paralysis by analysis," as industries can litigate against DTSC over unfavorable department decisions. Activists say California was poised to lead the way on toxics regulation but now is faced with potentially one of the weakest chemical-regulatory mechanisms in the nation. According to CHANGE (Californians for a Healthy & Green Economy), the revised regulation is a betrayal of the Green Chemistry promise and ignores two years of public input, while caving to backroom industry lobbying. Furthermore, it is a betrayal to public interest groups, businesses, and residents of California and legislators who supported the intent of this bill, to protect Californians and spur a healthy, innovative green economy. Environmentalists say the toxics department gutted the initiative at the behest of the chemical industry, and then put out the changes for public comment during a 15-day period just before Thanksgiving. This was a violation of the law requiring a 45-day public comment period when a substantial reworking of state regulations is proposed. The new Director of California's Department of Toxic Substance Control, Debbie Raphael, announced that mid-October 2011 is the new target date for new draft regulations to implement California's Green Chemistry Law and new draft guidelines were issued October 31, 2011. The public comment period for the latest version of the draft regulations ends December 30, 2011.

Implementation of CGCI has been delayed indefinitely beyond the January 1, 2011 deadline due to issues that arose after public review of the third draft. The third draft, which was made public December 2010, contains substantial revisions, including scaled back manufacturer and retailer compliance requirements that were not well received by the environmental community. DTSCs newest draft has made the following changes:

- All references of nanotechnology are excluded (nano referring to materials with dimensions of 1,000 nanometers or smaller); this change is significant because it would have been considered the most significant attempt to regulate nanomaterials based on environmental or health impacts.
- The new draft redefines "responsible entities," which originally referred to the entire business chain of consumer products distribution, including manufacturers, brand name owners, importers, distributors, and retailers, "responsible entities is now limited to manufacturers and retailers .
- DTSC prioritizes Children's products, personal care products and household products until 2016, after that point all consumer products.
- The new proposed regulations also eliminate the requirement that the DTSC develop a list of chemicals of consideration and products under consideration.
- New timeline for implementation of regulations

Condensation Reaction

A condensation reaction, is a chemical reaction in which two molecules or moieties, often functional groups, combine to form a larger molecule, together with the loss of a small molecule. Possible small molecules that are lost include water, hydrogen chloride, methanol, or acetic acid, but most commonly in a biological reaction it is water.

When two separate molecules react, the condensation is termed intermolecular. A simple example is the condensation of two amino acids to form the peptide bond characteristic of proteins. This reaction example is the opposite of hydrolysis, which splits a chemical entity into two parts through the action of the polar water molecule, which itself splits into hydroxide and hydrogen ions. Hence energy is required to form chemical bonds via condensation.

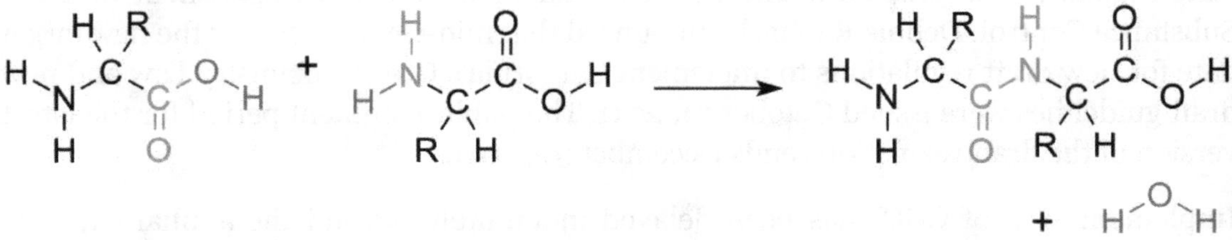

The condensation of two amino acids to form a peptide bond (red) with expulsion of water (blue)

If the union is between atoms or groups of the same molecule, the reaction is termed intramolecular condensation, and in many cases leads to ring formation. An example is the Dieckmann condensation, in which the two ester groups of a single diester molecule react with each other to lose a small alcohol molecule and form a β-ketoester product.

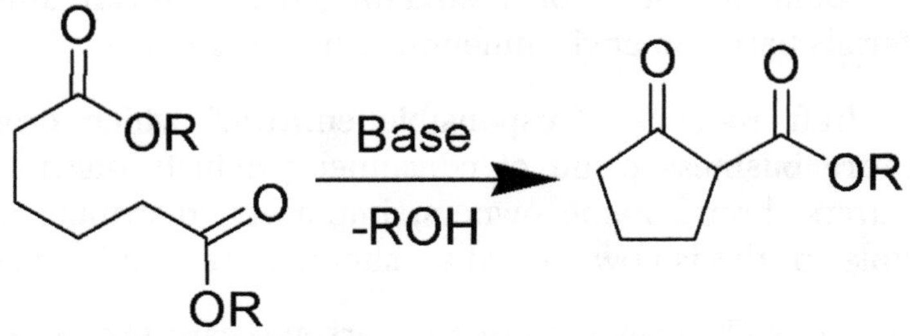

Dieckmann condensation reaction

Mechanism

Many condensation reactions follow a nucleophilic acyl substitution or an aldol condensation reaction mechanism. Other condensations, such as the acyloin condensation are triggered by radical or single electron transfer conditions.

Condensation Reactions in Polymer Chemistry

In one type of polymerization reaction, a series of condensation steps take place whereby monomers or monomer chains add to each other to form longer chains called polymers. This is termed 'condensation polymerization' or 'step-growth polymerization', and occurs for example in the synthesis of polyesters or nylons. It may be either a homopolymerization of a single monomer A-B with two different end groups that condense or a copolymerization of two co-monomers A-A and B-B. Small molecules are usually liberated in these condensation steps, in contrast to polyaddition reactions with no liberation of small molecules.

In general, condensation polymers form more slowly than addition polymers, often requiring heat. They are generally lower in molecular weight. Monomers are consumed early in the reaction; the terminal functional groups remain active throughout and short chains combine to form longer chains. A high conversion rate is required to achieve high molecular weights as per Carothers' equation.

Bifunctional monomers lead to linear chains, and therefore thermoplastic polymers, but, when the monomer functionality exceeds two, the product is a branched chain that may yield a thermoset polymer.

Applications

This type of reaction is used as a basis for the making of many important polymers, for example: nylon, polyester, and other condensation polymers and various epoxies. It is also the basis for the laboratory formation of silicates and polyphosphates. The reactions that form acid anhydrides from their constituent acids are typically condensation reactions.

Many biological transformations are condensation reactions. Polypeptide synthesis, polyketide synthesis, terpene syntheses, phosphorylation, and glycosylations are a few examples of this type of reaction. A large number of such reactions are used in synthetic organic chemistry. Other examples include:

- Acyloin condensation
- Aldol condensation
- Benzoin condensation (this is not technically a condensation, but is called so for historical reasons)
- Claisen condensation
- Claisen–Schmidt condensation
- Darzens condensation (glycidic ester condensation)
- Dieckmann condensation
- Guareschi–Thorpe condensation
- Knoevenagel condensation
- Michael condensation
- Pechmann condensation
- Rap–Stoermer condensation
- Self-condensation or symmetrical aldol condensation
- Ziegler condensation

References

- Sheldon, R. A.; Arends, I. W. C. E.; Hanefeld, U. (2007). "Green Chemistry and Catalysis". doi:10.1002/9783527611003. ISBN 9783527611003.
- Anastas, Paul T.; Warner, John C. (1998). Green chemistry: theory and practice. Oxford [England]; New York: Oxford University Press. ISBN 9780198502340.
- California Department of Toxic Substances Control. "What is the Safer Consumer Products (SCP) Program?". Retrieved 5 September 2015.
- Anastas, P.T., Levy, I.J., Parent, K.E., eds. (2009). Green Chemistry Education: Changing the Course of Chemistry. ACS Symposium Series. 1011. Washington, DC: American Chemical Society. doi:10.1021/bk-2009-1011. ISBN 978-0-8412-7447-1.
- "Green & Sustainable Chemistry Network, Japan". Green & Sustainable Chemistry Network. Retrieved 2006-08-04.
- "2005 Crystal Faraday Green Chemical Technology Awards". Green Chemistry Network. Retrieved 2006-08-04.
- "The Presidential Green Chemistry Awards". United States Environmental Protection Agency. Retrieved 2006-07-31.
- Thompson Hine Environmental Law Group,California Postpones Adoption of Proposed Green Chemistry Initiative, Retrieved 03/2011.
- California Department of Toxic Substances Control, California Green Chemistry Initiative, Phase Two: Recommendations Report, Retrieved 04/2011.
- California Department of Toxic Substance Control, Final Report of Green Chemistry Initiative, Retrieved 03/2011 from.

Permissions

All chapters in this book are published with permission under the Creative Commons Attribution Share Alike License or equivalent. Every chapter published in this book has been scrutinized by our experts. Their significance has been extensively debated. The topics covered herein carry significant information for a comprehensive understanding. They may even be implemented as practical applications or may be referred to as a beginning point for further studies.

We would like to thank the editorial team for lending their expertise to make the book truly unique. They have played a crucial role in the development of this book. Without their invaluable contributions this book wouldn't have been possible. They have made vital efforts to compile up to date information on the varied aspects of this subject to make this book a valuable addition to the collection of many professionals and students.

This book was conceptualized with the vision of imparting up-to-date and integrated information in this field. To ensure the same, a matchless editorial board was set up. Every individual on the board went through rigorous rounds of assessment to prove their worth. After which they invested a large part of their time researching and compiling the most relevant data for our readers.

The editorial board has been involved in producing this book since its inception. They have spent rigorous hours researching and exploring the diverse topics which have resulted in the successful publishing of this book. They have passed on their knowledge of decades through this book. To expedite this challenging task, the publisher supported the team at every step. A small team of assistant editors was also appointed to further simplify the editing procedure and attain best results for the readers.

Apart from the editorial board, the designing team has also invested a significant amount of their time in understanding the subject and creating the most relevant covers. They scrutinized every image to scout for the most suitable representation of the subject and create an appropriate cover for the book.

The publishing team has been an ardent support to the editorial, designing and production team. Their endless efforts to recruit the best for this project, has resulted in the accomplishment of this book. They are a veteran in the field of academics and their pool of knowledge is as vast as their experience in printing. Their expertise and guidance has proved useful at every step. Their uncompromising quality standards have made this book an exceptional effort. Their encouragement from time to time has been an inspiration for everyone.

The publisher and the editorial board hope that this book will prove to be a valuable piece of knowledge for students, practitioners and scholars across the globe.

Index

www.ingramcontent.com/pod-product-compliance
Lightning Source LLC
Jackson TN
JSHW052156130125
77033JS00004B/182
9781635491104